13 71 ⎯ 1 =

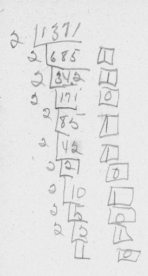

2 | 1371
2 | 685   □ □
2 | 342   □ □
2 | 171   □ 0
2 | 85   1 □
2 | 42   1
2 | 21   □ 0
2 | 10   □ □
2 | 5   □ □
2 | 2   □ □
   1   □

ELEMENTARY CONCEPTS OF MATHEMATICS

THE MACMILLAN COMPANY
NEW YORK · BOSTON · CHICAGO · DALLAS
ATLANTA · SAN FRANCISCO

MACMILLAN AND CO., LIMITED
LONDON · BOMBAY · CALCUTTA · MADRAS
MELBOURNE

THE MACMILLAN COMPANY
OF CANADA, LIMITED
TORONTO

ELEMENTARY CONCEPTS
OF Mathematics

Burton W. Jones
PROFESSOR OF MATHEMATICS, CORNELL UNIVERSITY

1947

THE MACMILLAN COMPANY · NEW YORK

TO
MY PARENTS

Preface

In 1939 it became evident to members of the mathematics
department of Cornell University that there was need for a
course designed for students who have had a minimum of
training, who do not expect to take other courses in the subject, but who want a firmer grounding in what useful mathematics they have had and wish such additional training as
they, as nonmathematicians, may find interesting and/or
useful in later life. Toward this end several members of the
department cooperated: R. P. Agnew wrote a chapter on
rational and real numbers, F. A. Ficken one on graphs,
Mark Kac one on probability, Wallace Givens chapters on
"mirror" and Lorentz geometries, R. L. Walker one on
topology, and this author chapters on integers and algebra.
With the advice and criticism of W. B. Carver and W. A.
Hurwitz, the book was edited and prepared for lithoprinting
by Wallace Givens and the author and appeared in that
form in 1940 under the authorship of "the department of
mathematics." From its beginning the course has had a
clientele from all over the university and seems to have been
growing in favor with students and faculty. In the light of
experience the text has been thoroughly revised by the author
and numerous additions made, including a chapter on logic
in the preparation of which he wishes to acknowledge the
assistance of his father, Arthur J. Jones.

While this text is designed for college students, almost no
previous training is assumed beyond an acquaintance with
integers and fractions and, for Chapters VII, VIII, IX, a
little plane geometry.

First, much space and effort have been given to an attempt to cultivate an *understanding* of the material. To this end stress is laid on the distributive law for numbers, since it and the use of parentheses is at the root of much difficulty in algebra. Letters are used almost from the beginning to stand for numbers so that the student may become used to the idea. At times the subject is admittedly made harder by efforts to make it understood. For instance, in the treatment of annuities the student would more quickly solve the problems at hand if formulas were developed for annuities certain and various types of problems classified, but he is more likely to be led to an understanding of the subject if he thinks through each problem on its own merits. Second, an attempt is made to straighten out in the student's mind certain mathematical concepts of his everyday life which are usually only dimly understood: compound interest, the graph, averages, probability and games of chance, cause versus coincidence. Third, there is an effort to cultivate *appreciation* of mathematics — not awe of mathematics which he can never understand, but a joy in it and a respect for its methods. From this point of view, algebra is chiefly a tool for finding formulas — a method of expressing certain things clearly and succinctly so that, after the process is gone through with, the rest is mere substitution. It is hoped that the student will feel that a proof is a way of showing something rather than an exercise imposed by the teacher. Numerous references to other books are given in an effort to encourage the student to study some topic further and/or to acquire a little acquaintance with mathematical literature. In fact, in teaching the course the author has often required a term paper on some one of the suggested topics. Three of the references given (the books of Bell, of Hardy, and the article by Weyl) are especially useful in showing how mathematicians think about themselves. Much puzzle material is introduced — this for a threefold reason: to make the material enjoyable, to provide a means for recreation now and later, to remove some of the undeserved mystery clinging about puzzles mathematical. Fourth, there is much emphasis on *logical development*. Not only in the chapter on logic but throughout the book an effort is made to cultivate logical statement and reasoning. Not many facts are stated without at least an indication of the proof. Fifth, the book is written in the belief that it should be a textbook, not a novel, and that a pencil and paper and the working of exercises are, if anything, more necessary to an understanding of mathematics than is work in a laboratory to the understanding of science. Finally, the book contains much material useful for the prospective teacher in secondary schools. In fact, the lithoprinted edition has seen such use.

Chapters VII, VIII, and IX are introduced to show the student that there are geometries he never dreamed of, to show how certain facts can logically be found, to provide puzzle material, and to encourage the student to do some investigating on his own. It is interesting to find how much easier topics like groups and topology are for students than algebra which has been learned but poorly. They find these subjects interesting and the proofs understandable.

Teachers may notice the omission of any mention of trigonometry and calculus. The author feels that students who are not going on to further mathematics or science would find little use for these subjects.

While stress is laid on the interrelations of the subjects in this book, there are numerous ways in which the order and emphasis may be changed. In fact, it has been the author's experience that even if two chapters are omitted, there is too much in the book for a three-hour course for two semesters. Topics which can be omitted without affecting the treatment of others, except in some cases to force selection of exercises, are given in the table on page x. If "numbers on a circle" are omitted, then tests for divisibility, groups, and Diophantine equations must also be left out. Some teachers may prefer to take up the chapter on logic at the end of the course.

Topic	Chapter	Sections
Logic	I	Entire chapter
Numbers to various bases and Nim	II	6, 7, 8, 9
Divisors and prime numbers	II	13, 14
Groups	II	15
Annuities	IV	5
Nonterminating progressions	IV	7
Antifreeze formula	IV	9
Puzzle problems	IV	10
Puzzle problems and Diophantine equations	IV	10, 11
Find a formula which fits or almost fits	V	7
Mirror geometry	VII	Entire chapter
Lorentz geometry	VIII	Entire chapter
Topology	IX	Entire chapter
Five-color theorem	IX	4

The author wishes to express his appreciation of the work of the Macmillan Company in the preparation of this book and for their sympathetic counsel. As may be seen from the history of this book, several persons beside the author had to do with its inception, and without them and that abstract entity, the department of mathematics, this book would not have been. May they and it not be ashamed of the outcome!

Burton W. Jones

Ithaca, N. Y.

Contents

ELEMENTARY CONCEPTS OF MATHEMATICS

CHAPTER I

Logic

> "Contrariwise," continued Tweedledee, "if it was
> so, it might be; and if were so, it would be; but as it
> isn't, it ain't. That's logic."

1. Statements.

The tools of a carpenter are his hammer, saw, plane, rule,
and all the contents of his tool kit; the tools of a chemist
are his chemicals, Bunsen burner, test tubes, and so forth;
the tools of a mathematician are books, a pencil, paper, and
logic.[1] Just as it behooves a carpenter to learn something
about his hammer before he sets himself up as a carpenter
even though carpentry will increase his proficiency in the
use of this tool, so anyone about to do some mathematics
should be somewhat familiar with logic before he begins,
even though doing mathematics should give him practice in
the use of logic.

Thus, we propose in this beginning chapter to look into
the structure of logic. In order to make this examination as
clear as possible, we shall at first consider logic as it appears
outside of mathematics. In the later chapters of this book
we shall learn more about logic as it is used in mathematics.

An important part of any logical reasoning is the statements of which it is composed. We shall here not be concerned too much with the truth or falsity of statements, for
this is a matter of experience. But the exact meaning of a
statement we should examine in detail.

---

[1] L. E. Dickson maintained that the most essential tool of a mathematician
is a large wastebasket.

One of the most difficult things to do is to say exactly what you mean. It is even harder to understand what the other fellow means, especially when you are not sure he is speaking precisely. It is no wonder that there are so many misunderstandings and acrimonious arguments. In fact, one might guess that many disagreements are fundamentally misunderstandings. Though it is possible for two persons to understand each other when one does not say what he means nor the other correctly interpret what the first says, such a situation is like making two mistakes in a problem and coming out with the right answer. The best way to communicate one's thoughts accurately is to state them precisely in the hope that they will be understood. In phases of certain games, like bidding in bridge, one of the chief objects is communication from one partner to the other. In fact, it is hard to find any phase of our life that does not depend on accurate communication by written or spoken word. Mathematics is no exception to this statement. In this subject, however, precision of statement and proof is easier than elsewhere. For this reason it is desirable in this chapter to give examples largely outside of mathematics.

Precisely what is meant by the statement:

1. If plants become green, they must have had sunshine.

One way to explain it is to say the same thing in different ways: all plants must have sunshine to become green; in order to become green, it is necessary for plants to have had sunshine; all plants which become green are included among those plants which have had sunshine. Still another way to make the statement is to interchange the two parts and say: plants must have had sunshine if they become green. But it is quite another thing to put the "if" in another place and say:

2. If plants have had sunshine, they become green.

Two other statements may be formed from sentences 1 and 2 by inserting the word "not" in each part:

3. If plants do not become green, they must not have had sunshine.

4. If plants have not had sunshine, they do not become green.

For such statements as these, a certain type of diagram is useful. We draw two concentric [1] circles as in Fig. 1:1 and label the outer one A and the inner one B. We let A include all plants which have had sunshine and B all plants which become green. The diagram then represents the statement that the category of all green plants is entirely within (that is, is smaller than) that of plants which have had sunshine. This is **equivalent** to statement 1, that is, says the same thing in different words — each implies the other. The connection between statements 1 and 2 can be seen by noting that the latter says that anything in circle A is in circle B; this does not necessarily follow. *Statements 1 and 2 are not equivalent.* To evaluate the other two statements, notice that in the diagram the outside of circle A represents plants which have not had sunshine, and the outside of B plants which do not become green. Statement 3 says that anything outside B is outside A; this again does not follow from the diagram. On the other hand, statement 4 says that anything outside A is outside B which does follow from the diagram. Furthermore, if everything outside A is outside B, it follows that everything inside B is inside A. *Thus statements 1 and 4 are equivalent but neither 2 nor 3 have the same meaning as 1.*

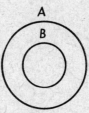

Fig. 1:1

Notice that statement 4 can be obtained mechanically from 1 by interchanging the two parts of the statement and inserting the word "not" in each part. Each is called the **contrapositive** of the other. Similarly statement 3 is the contrapositive of 2 and hence 2 and 3 are equivalent statements. They are, in fact, two forms of the converse to statement 1 or of statement 4.

---

[1] This restriction, "concentric," is merely a means of making one circle lie entirely within the other.

The diagram given is not by any means the only one which fits the original statement. We might draw the same circles but let the *outside* of circle B represent plants which have had sunshine and the outside of A plants which become green, the latter being the smaller category. Then the inside of B would represent plants which have not had sunshine and the inside of A those which do not become green. This diagram would equally well apply to the comparison of the four statements.

However, one must be careful in the labeling of the circles. For instance, we might be tempted to use two circles in which A represents all the things which green plants must have, and B is sunshine. This represents the bare statement but is apt to be misleading in its implications. Points inside A and outside B then represent things besides sunshine which green plants must have and quite properly so. But one is apt to misread this diagram and conclude that if a plant does not have sunshine it still may be green, which contradicts the statement made. Another objection to such a labeling of this diagram is that it might be interpreted to mean that green plants must have all the constituents of sunshine. One should always check a diagram by asking himself: is everything in the category represented by the smaller circle within the category of the larger circle?

Moreover, to label one circle Green Plants, and the other Sunshine, is to invite fruitless argument as to whether plants are in the sunshine or sunshine in the plants.

It is usually safer to think of the contents of the circles as objects, things, or sets of objects or things. For instance, the statement "whenever it snows he stays home from church" can be diagrammed more easily if replaced by the more awkward "Every snowy day is a stay-at-home-from-church-day for him."

2. Always, never, and sometimes.

Many statements include explicitly or implicitly one of these three words or their close relatives: "all," "none," or "some." The first two words are very stringent ones and a statement preceded by one of these adverbs is likely to be very hard to establish. We often find ourselves in the position of Captain Corcoran of the "King's Navy" whose classic reply when challenged was, "Well, hardly ever." The reason, of course, that a statement involving "never" or "always" is difficult to make with certainty is that *one single exception suffices to prove it false.* If I made a statement that it never rains in Death Valley, it would be proved false if the oldest inhabitant could prove that one day when he was a boy it did rain. True statements involving one of these words are easier to find in mathematics than elsewhere, probably because the situation is here better controlled. I can, for example, say that no matter what number you give me I can *always* mention a number larger; this, of course, is due to the fact that any number is increased if you add 1 to it. This and many other such statements are more complex examples of such an assertion as: it is *always* true that if anything is white it is white.

It will probably be admitted that most of our "always" and "never" statements in everyday life are, and are known to be, exaggerations. They are working hypotheses which are assumed innocent until they are found guilty. You ask your neighbor, "Won't you keep my dog during my absence?" "Oh no," he replies, "he chases cats and I don't want mine chased." You say, "He doesn't chase my cat — so he won't chase yours." Just as your argument had holes in it, so had his, even though he may have seen your dog chasing a cat. It is rather a case of weighing the evidence and drawing conclusions as to probable future behavior.

These words ("always," "never," etc.) often occur implicitly. In the statement used for illustration in section 1 was implicit the idea that *all* plants which become green have had sunshine. This statement would be shown false if one could give *one* example of a plant which became green without having had sunshine. Such an example would disprove 4 but not 2 nor 3.

When the word "some" is used, the diagram contains overlapping circles. Suppose we have the statement: some men are white. This statement has the diagram of Fig. 1:2

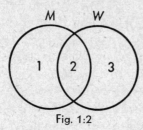

Fig. 1:2

where M includes all men and W all white animals. Portion 1 includes all men who are not white, portion 2 all white men, and 3 all white animals not men, while the outside of the two circles represents all things (or animals) which are neither white nor men. Of course, it might be that a proper diagram should be one in which one circle is entirely within the other, but this would not be possible unless a stronger statement than the given one could be made.

One can represent the word "most" in a diagram by considering the relative size of the common portion. For instance, if the statement were: most men are white animals, the larger part of circle M should be made to lie within W. However, since the statement does not imply that most white animals are men, one could not tell whether the larger part of circle W should lie in M. That would depend on the relative size of M and W. If there were fewer white animals than men, then most white animals would be men, but if there were more white animals than men, one could not tell whether the larger part of W were in M unless one had further information such as, for instance, that W was at least twice as large as M. On the other hand, notice that the statement: *some men are white animals* is equivalent to *some white animals are men*.

EXERCISES

In each case below, a statement is given followed by several others. Tell which of these statements are equivalent to the given one and which are not. Draw a diagram for the given statement. What kind of example, if any, would suffice to show the given statement false? Express the given statement in other words.

1. All Polynesians are brown.
   a. If a man is Polynesian, he is brown.

   b. If a man is not brown, he is Polynesian.

   c. In order to be brown, a man must be Polynesian.

   d. Whenever a man is Polynesian, he is sure to be brown.

   e. If a man is not brown, he cannot be Polynesian.

   f. If a man is not Polynesian, he is not brown.

2. All Americans are Canadians.

   a. Every Canadian is an American.

   b. If a man is an American, he must be a Canadian.

   c. If a man is not an American, he is not a Canadian.

   d. If a man is not a Canadian, he is surely not an American.

3. When the sunrise is red, it is sure to rain during the day.

   a. If it is raining today, then the sunrise must have been red this morning.

   b. If it does not rain today, then the sunrise must have been red.

   c. If it does not rain today, then the sunrise must not have been red.

   d. Whenever it rains during the day, it began with a red sunrise.

4. If food is sweet, it has honey in it.

   a. Honey is sweet.

   b. If food does not have honey in it, it is not sweet.

   c. If food is not sweet, it does not have honey in it.

   d. Sweet food always contains honey.

5. It never rains in the summer.

   a. If it is summer, it is not raining.

   b. If it is not raining, it is not summer.

   c. In the summer it never rains.

   d. Never in the summer does it rain.

   e. If it is raining, it is not summer.

   f. Sometimes in the summer, it does not rain.

6. Some Nazis are cruel.

   a. He is a Nazi. Hence he is cruel.

   b. He is cruel. Hence he is a Nazi.

   c. He is not a Nazi. Hence he is not cruel.

   d. He is not cruel. Hence he is not a Nazi.

7. Which of the following statements are false and which are true? Prove your answers are correct or describe a method of establishing your conclusions:

   a. Every even number is divisible by 4.

   b. It never rains while the sky is cloudless.

   c. Every even number is divisible by 2.

   d. A banker always has money in his pocket.

   e. Needles and pins, needles and pins,
   When a man marries his trouble begins.

3. Arguments.[1]

When we bring together several statements and from them deduce a new one whose validity depends solely on the validity of the original statements, we have what we call an **argument.** The original statements are called **premises** or **hypotheses,** the resulting one a **conclusion,** and the connecting link a **deduction.** There are two principal uses to which arguments are put. In the first place, we may wish to convince a friend (or perhaps ourselves) that something we say is true. Or we may try, from known facts, to gain further information which we do not already know or perhaps even suspect. In the former case the conclusion is known, but it is not to be assumed, and, especially in the case of a recalcitrant friend, it is often desirable to conceal it. Our task is first to persuade him to accept some hypotheses and then, by inexorable logic, force him to the conclusion; once he accepts the hypotheses, he is lost. The situation is similar to that when one approaches a modern toll bridge — from a certain point on there is no turning back. In fact, all of our reasoning must be of this character except where we lay ourselves open to the criticism of James Harvey Robinson in his *Mind in the Making:* "most of our so-called reasoning consists in finding arguments for going on believing as we already do."

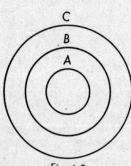

Fig. 1:3

Once our friend is so careless as to admit that (1) all Pennsylvanians are Americans, and (2) all Americans are kind to animals; then he is forced to agree that all Pennsylvanians are kind to animals. In this connection diagrams for simple arguments are again useful. We now draw three concentric circles: A, B, C. The first statement has as a diagram the two smaller circles where A includes all Pennsyl-

---

[1] Cf. reference **31,** Chap. II. Boldface numbers refer to the Bibliography at the back of the book.

vanians and *B* all Americans. The second statement has the two larger circles as a diagram where *C* stands for all persons who are kind to animals. We know that *if A is inside B and B is inside C, then A is inside C.* The name for this property of insideness is **transitivity**. This is shown in the diagram by ignoring the circle *B*.

There is sometimes a tendency to reverse this diagram which, unless care is used, leads to wrong results. One could let the largest circle embrace all the characteristics of Pennsylvanians, circle *B* all the characteristics of Americans, and *A* all the characteristics of those kind to animals. One would properly deduce that among the characteristics of Pennsylvanians are all the characteristics of those who are kind to animals and hence all are kind to animals. But to label the outside circle Pennsylvanians and the next one Americans would be likely to lead one to think that the latter was the smaller category, contrary to the original statement.

Sometimes, in an argument, certain very obvious hypotheses are not stated. For instance, if a man told you, "I would rather see my daughter in the grave than see her married to you," you would probably conclude that he did not fancy you as a prospective son-in-law. The obvious and unmentioned hypothesis is that the stern parent would be very sorry to see his daughter in the grave. However, this hypothesis is very necessary to the argument. Another example of this point is the usual child's conclusion from a statement like "if you don't stop crying I shall send you to bed." The fact that he concludes from this that if he does stop crying you will not send him to bed is not, at least for an intelligent child, a mere confusion of a statement with its converse, but a recognition of the fact that if you did not mean the converse as well as the statement, that is, if he would be sent to bed anyway, there would be no use in stopping his howl.

As was mentioned above, the use of an argument is to establish certain conclusions. A true conclusion does not imply a valid argument nor a true hypothesis. In this connection the following table is enlightening:

Hypotheses	T	T*	T	T	F	F	F	F
Deduction	V	V	NV	NV	V	V	NV	NV
Conclusion	T	F	T	F	T	F	T	F

where T stands for True, F for False, V for Valid, and NV for Nonvalid. Except for the second, one can give an example of each of the eight possibilities. For instance, for the following we have the third possibility, that is, the hypotheses and conclusion are true but the deduction is not valid: all men are human, all dogs are mammals; hence all men are mammals. It is easy to find an example in which the hypothesis is true, the deduction not valid, and the conclusion false. After watering the lawn on a clear day we could have the following argument in which one hypothesis would be false, the deduction valid, and the conclusion true: it is raining and whenever it rains, the grass in the open is wet; hence the grass is wet. Examples for the remaining three possibilities are left as exercises.

The above table then shows that, knowing the truth or validity of two of the three ingredients: hypothesis, deduction, conclusion, tells nothing about the truth or validity of the third with the following exceptions.

1. If the hypotheses are true and the deduction valid, the conclusion must be true.
2. If the conclusion is false, either an hypothesis must be false or the deduction not valid or both.

These two statements are equivalent. The latter will be recognized as the basis for the "indirect proof" or *reductio ad absurdum* argument of our high-school geometry. An example of it is the following: "I know John did not leave the house because if he had, there would have been tracks in the snow. But there are no tracks. That proves it." The argument is impeccable, the conclusion false, and hence the supposition that John left the house is untenable.

A more complex example of the "indirect proof" is the

following. "Is it true that all men are brutes? If all men were brutes, then none would be good nurses. But I know a man who is a good nurse. Hence it is not true that all men are brutes." The hypotheses, one of which we wish to prove false are: (1) all men are brutes, and (2) no good nurses are brutes. Our picture would then be Fig. 1:4 where

M includes all men, B all brutes, and G all good nurses. From our hypotheses we then have the conclusion that no men are good nurses. A *single* exception suffices to show this conclusion false. The argument has no flaw and hence one of the hypotheses is false.

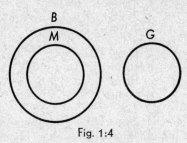

Fig. 1:4

We accept 2 as true and thus 1 is false. Notice that if hypothesis 1 were "some men are brutes" we could not show it false by the above means. We could not tell from the given information whether or not a portion of M lay in G.

EXERCISES

1. Give the three examples "left as exercises" in the previous section.

2. What conclusions could be drawn from the two following hypotheses? Discuss the various possibilities with regard to the overlapping circles: some men are white animals, some white animals are Americans.

3. Draw a diagram showing the relationships among the three things: seagoing vessels, implements of war, submarines. What statements could be made on the basis of your diagram? Make the list as complete as you can.

4. Do the same as in the above diagram, with "submarines" replaced by "airplanes."

For each of the following arguments, state which are hypotheses and which are conclusions. State which conclusions follow and which do not. Except for exercises 17, 25, 26, 27, draw diagrams. If you are not sure exactly what some of the statements mean, state carefully your interpretation of their meaning. If there are any unmentioned hypotheses, point them out.

5. All Polynesians are brown. All Balinese are Polynesians.
  a. Hence all Balinese are brown.
  b. Hence all brown people are Polynesians.

c. Hence all brown people are Balinese.

d. If a man is not Balinese, he is not brown.

e. If a man is not brown, he is not Balinese.

f. Statements like *d* and *e* involving *Polynesian* and *brown.*

g. Statements like *d* and *e* involving *Polynesian* and *Balinese.*

6. All rational numbers are real. If a number is an integer, it is a rational number.

a. Hence if a number is a rational number it is an integer.

b. Hence if a number is an integer, it is real.

c. Hence all real numbers are rational.

d. Statements as in exercise 5, parts *d* to *g* with "rational number," "real number," "integer" taking the places of "Polynesian," "brown," "Balinese" respectively.

7. All Americans are Canadians. If a man is a Mexican, he is an American.

a. Hence all Mexicans are Canadians.

b. Hence if a man is a Canadian, he is a Mexican.

c. Some Canadians are Mexicans.

d. Statements analogous to exercise 6*d.*

8. If any food is sweet, it has honey in it. All honey contains sugar.

a. Hence sugar is sweet.

b. Hence any food containing sugar is sweet.

c. Hence any food which is sweet contains sugar.

d. Hence honey is sweet.

e. Statements obtained by filling in the following blanks in six possible ways with the three words: "honey," "sugar," "sweet." "If something is not ____, it is not ____."

9. Whenever it is going to rain my rheumatism bothers me.

a. My rheumatism bothers me. Hence it is going to rain.

b. My rheumatism does not bother me. Hence it is not going to rain.

c. It is not raining. Hence my rheumatism does not bother me.

10. If you wear your sweater near that bull, he will gore you for your sweater is red and anything red infuriates him.

11. I know I won't like Brussels sprouts for they are just like little cabbages, and I don't like cabbage.

12. Father says, "Never do today what you can put off until tomorrow," and I act on everything father says. Hence the only things I shall do tomorrow are the things I don't need to do today.

13. All boys are bad. Most bad persons go to jail. Hence,

a. Some boys go to jail.

b. Some persons who go to jail are boys.

c. Not all persons who go to jail are boys.

  d. Of those who do go to jail, some are boys and some not.

  e. Some who go to jail are bad.

  f. Most persons who go to jail are bad.

  g. Most of those who go to jail are bad boys.

 14. "It is easier for a camel to go through a needle's eye than for a rich man to enter into the kingdom of God."

  a. Hence if I am poor, my chances of entering His kingdom are good.

  b. Hence if I want to enter His kingdom easily, I must be poor.

 15. "Blessed are ye when men shall revile you, and persecute you, and say all manner of evil against you falsely, for my sake. Rejoice, and be exceeding glad: for great is your reward in heaven: for so persecuted they the prophets which were before you." I am being persecuted. Therefore I am to have a great reward.

 16. "If any man desire to be first, the same shall be last of all, and servant of all." Now I serve everybody, hence I shall be first.

 17. "Let us start with these propositions:

  (1) It is the duty of our Government not to take any action which will diminish the opportunities for the profitable employment of the citizens of the United States. . . .

  (2) It is not in the interest of the nation to adopt any policy which makes the United States, in peace or in war, needlessly dependent upon the will of any foreign nation for any essential supply. . . . Applying these principles to foreign trade and considering the nation as a unit, we do a useful thing if we exchange one hour of foreign labor, only if the one hour of foreign labor could not have been performed by an American. We do not, as is sometimes imagined, increase our wealth if we exchange one hour of American labor for two or more hours of foreign labor. That looks like a bargain, but it is a bad bargain, for not only do we deprive the American of his opportunity to work, but also we withdraw from the workman affected a purchasing power which reacts to the benefit of other sections of industry and of agriculture generally. Also we may weaken our self-reliance." — G. N. Peek, *Why Quit Our Own*, p. 213. D. Van Nostrand Co., New York, 1936.

 18. If you cannot do mathematics, you cannot be an engineer.

  a. Hence if you wish to be an engineer, you must learn mathematics.

  b. Hence if you learn mathematics, you can be a good engineer.

 19. I know he doesn't love me for men never tell the truth to a girl, and he told me he loved me.

 20. I know he loves me for men never tell the truth to a girl, and he has not told me he loved me.

21. I know he loves me for men never tell the truth, and he told me he does not love me.

22. I know he doesn't love me for men in love never tell the truth, and he told me he loved me.

23. The moon is made of green cheese. All cheese has holes in it. Therefore the moon has holes in it.

24. I would not dare throw a spitball in class because the teacher would be sure to see it. Then she would write a note to my mother and my father would tan my hide.

25. "If the frame has been welded or straightened, most purchasers would do well not to consider the car further, for a car that shows evidence of having been in a wreck represents a pretty doubtful purchase, however low its price may seem." — *Consumers' Research*, February, 1942.

26. *Antony:*　　　　"The noble Brutus
　　　　　　　　Hath told you Caesar was ambitious;

　　　　　　　.　　.　　.　　.

　　　　　　　You all did see that on the Lupercal
　　　　　　　I thrice presented him a kingly crown,
　　　　　　　Which he did thrice refuse. Was this ambition?
　　　　　　　Yet Brutus says he was ambitious;
　　　　　　　And, sure, he is an honorable man."
　　　　　　　　　　　　　　— *Julius Caesar*, III, ii.

27. *Polonius:* "Neither a borrower nor a lender be;
　　　　　　　For loan oft loses both itself and friend,
　　　　　　　And borrowing dulls the edge of husbandry.
　　　　　　　This above all: to thine own self be true,
　　　　　　　And it must follow, as the night the day,
　　　　　　　Thou canst not then be false to any man."
　　　　　　　　　　　　　　— *Hamlet*, I, iii.

4. Causality.

It is one of the axioms of numbers that two things equal to the same thing are equal to each other. But that does not imply that two things which contain the same thing contain each other, or that two things which cause the same thing cause each other. As in the case of arguments, causality is much of the time a one-way road. A cold wind as well as a hot fire makes one's face glow, but the wind does not cause the fire nor the fire the wind. We could represent this situation by the diagram of Fig. 1:5 where W and F stand for the wind and fire and G stands for the glowing face. There need not be any causal connection between W

and *F*. Only if one of the arrows were reversed could we establish such a connection: if a hot fire were caused by a glowing face, then a cold wind would indeed cause a hot fire.

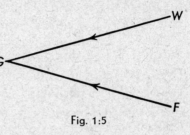

Fig. 1:5

A full moon makes the hounds bay the moon, and lovers' hearts beat quicker, but few would claim that the hounds baying the moon made lovers' hearts beat quicker or that the latter caused the hounds to bay at the moon. Here we have the diagram of Fig. 1:6 where *M*, *H*, and *L* stand for the moon, the hounds baying and the hearts beating, respectively. There is no causal connection between *H* and *L*. Moreover, one cannot even conclude that *when* the hounds bay, lovers' hearts beat quicker, since there may be many times when the hounds bay and the moon is not full.

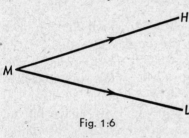

Fig. 1:6

One has a similar situation in the common argument that to increase your pay you should get a college education, for those who have such an education as a group get higher pay than those who do not. The conclusion may be correct, but the argument is not sound, for it is possible that getting a college education and getting high pay may both be the result of having parents of a higher class and with larger income; also, there is *some* evidence to the effect that only the more intelligent can get such an education. In other words, the following two statements are not in the least equivalent.

1. Those who go to college draw higher salaries than those who do not.
2. Those who have gone to college are drawing higher

salaries than they would have drawn had they not gone to college.

Furthermore, even if the latter statement held for college graduates as a whole, it might fail to hold for many individuals.

It is even possible for two things which occur together to have no common cause. For instance, my clock strikes twelve when the noonday whistle blows, but the clock strikes then because it is a good clock and I wind it, while the whistle blows because someone blows it.

If A always causes B it is true that if A occurs, then B must occur. Also, if A usually causes B it is true that when A occurs then B usually occurs. But the converse of neither of these statements is true. It is at this point that we have one of the commonest pitfalls of everyday reasoning. Even if hounds bayed the moon only when it was full, they might not cause lovers' hearts to beat quicker. The sun unblocked by clouds causes the day to be bright, and whenever the day is bright there is the sun unblocked by clouds; but no one would claim that the bright day causes the sun to shine. If one of a man's eyes is blue, it is usually true that the other is also, but the blueness of the one is not caused by the blueness of the other. You may have believed in your childhood that angleworms came down from the sky because after a rain they appeared in great numbers on the pavement. Of course, it is true that if the angleworms did come down in the rain, then they would appear under the conditions observed, but that lends no support to the superstition.

A cause is one of many reasons. The former, strictly speaking, does not often occur in mathematics for our reasons are likely to be in the nature of relationships or arguments. The answer to the question, "Why is $x(x + 1)$ always even when x is an integer?" is: if x is even it is even, if it is odd, $x + 1$ is even, and in both cases we have a product of two integers one of which is even; such a product is always even. Moreover outside of mathematics, our answer to "why?" is often to give a reason which is not connected with cause even though our answer begins with "because." We have, "Why is he brown? Because he is a Malayan, and all Malayans are brown."

It is very hard to pin down this causal relationship. We can agree that if A causes B, then whenever A occurs, B must occur, but we mean more than that. Usually cause precedes effect (it certainly never follows it), but many times it fails to have this property, e.g., the sun causing glare on the water. We have seen that if it is true that whenever A occurs B occurs, it does not follow that A causes B. However, in scientific practice such occurrence is assumed to be evidence of cause. For instance, to investigate the truth of the statement "sunshine makes plants become green" the scientist might take two identical plants, give one sunshine over a period of time and deprive the other of it but keep all other factors the same. If, at the end of the time, the former were greener than the latter, and if the results of many experiments were the same, he then would conclude that sunshine caused plants to become green. One could perform a similar experiment on individuals with regard to a college education. If a set of persons of the same background and ability (and here is the rub in the experiment) were divided into two parts and one sent to college and one not, then if the former part received higher pay than the latter, the causal relationship would be more definitely established.

EXERCISES

In each of exercises 1 through 7 below, point out what conclusions follow logically and what do not. Also mention possible common causes for two or more things which occur together.

1. The depression occurred during Hoover's administration.
   a. Hence he was one of those who brought it about.
   b. If we feel a depression coming on, we should see to it that he is not in office.
2. Countless charts show that prices have varied with the amount of gold in this country.

    a. Hence to regulate prices, regulate the gold content of the dollar.

    b. Hence to regulate the gold in the country, regulate prices.

3. In the last ten years our main roads have been made wider and wider. And accidents have doubled.

    a. Therefore we should make narrower roads.

    b. The reason for the increase in the accident rate is that on wider roads men drive faster. Hence if we made the roads narrower, men would not drive so fast, and there would be fewer accidents.

4. One of the causes for feeble-mindedness is excessive drinking, for statistics show that the two go together to a surprising degree.

5. It has been observed that during very stormy weather there are invariably sunspots.

    a. Hence sunspots cause the storms.

    b. Hence storms cause sunspots.

    c. Hence when there are sunspots there are usually storms.

    d. Hence when there are storms there are usually sunspots.

6. There is a larger proportion of criminals among those who have only elementary education than among those with more extended education. Therefore to reduce criminality, increase the age of compulsory attendance in school.

7. *Polonius:* "Costly thy habit as thy purse can buy,
        But not express'd in fancy; rich, not gaudy;
        For the apparel oft proclaims the man,
        And they in France of the best rank and station
        Are most select and generous in that."

                                   — *Hamlet*, I, iii.

8. John had never eaten yeast and on January 1 weighed 120 pounds. From then until April 1 he ate three yeast cakes a day. At the end of that time he weighed 150 pounds. Which of the following conclusions are justified and to what extent?

    a. Eating yeast made John gain weight.

    b. Eating yeast makes people gain weight.

What additional facts would bolster these conclusions?

9. Quote an advertisement appearing in the daily paper, magazine, or heard over the radio — an advertisement which has to do with cause and effect. Point out to what extent the conclusions are justified.

5. Logical structure.

The usual arguments which we meet in everyday life are more complex than those previously given and as a result it is more difficult and usually not very enlightening to reduce them to diagrams, though parts of an argument may often be clarified by such a reduction. As a matter of fact, many statements cannot be reduced to any diagram we have given; e.g., the statement: no matter what number you mention, I can always mention a bigger one. In more complex arguments, the logical structure is best analyzed by rephrasing it.

Consider the following:

"Although most men carry insurance continuously during most of their lifetimes, life insurance statistics indicate that the average period for which an insurance policy remains in force is less than ten years. This indicates that many insurance contracts are discontinued every year, probably because they are not suited to the needs of the policyholders. Obviously, if insurance is carried during the lifetime of an individual, the payment of several agents' commissions and other acquisition expenses at intervals of nine or ten years is costly. Before any insurance is purchased, therefore, it should be studied carefully in order to determine whether or not it is suitable and the best for the purpose." [1]

The argument might be rephrased as follows: 1. Although most men carry insurance continuously during most of their lifetimes, life insurance statistics indicate that the average period for which an insurance policy remains in force is less than ten years. 2. Hence, many men discontinue policies and take out new ones. 3. Under these circumstances extra commissions and other expenses are paid. 4. This is costly. 5. Hence, one way to avoid extra cost is to avoid situation 2 and hence to eliminate as many causes of the situation as possible. 6. Probably in many cases the cause is that the policy is not suitable. 7. One of the reasons for the unsuitability is lack of care in selecting a policy. 8. Hence one way to save extra expense is to use care in selecting a policy.

Statements 1 and 2 are not really part of the main argument; their purpose is to show that what follows is a frequent consideration. The logic seems sound but there are several by-arguments that apply. Statement 2 follows 1 because the span of manhood is more than ten years. State-

[1] From Harwood and Francis, *Life Insurance*, p. 102. American Institute for Economic Research, Cambridge, Mass., 1940.

ment 5 is almost axiomatic but stems from the fact that if
you find all the things that produce a result and eliminate
the things, you eliminate the result. Statement 6 derives
strength from the following: if a man found he could not
meet the premiums he would not be able to take out another
policy; neither would he do this if he did not approve of life
insurance; the only other cause would be the unsuitability
of the policy or the company. Notice that the argument
does not show that if one uses care in selecting the policy,
extra expense will necessarily be avoided — the authors do
not in fact claim that. Other factors, such as a change of
circumstances of the policyholder, may render a policy un-
suitable no matter what care was taken in selecting the
policy. It might even be that a change in circumstances
would make a policy selected with care less suitable than
one not so selected.

There is one kind of diagram which is perhaps useful in
this situation. We could represent the sequence of causes by

$$L \rightarrow U \rightarrow D \rightarrow E$$

where L is lack of care in selection, U is unsuitability of
policy, D is discontinuance of policy, E is expenses to be
paid. Statement 1 is used to show D occurs, and the conclusion is that a way to avoid E is to avoid L.

We have not mentioned some important factors which
play a vital role in everyday reasoning. One of these is
experience which tells us what is relevant and what is not —
how much and what kind of evidence one needs before arriving at a reasonably sure conclusion. For instance, one sees
the sign:

WE HAVE EATING SPACE FOR 150 PEOPLE
SO YOU CAN BE SURE OF PROMPT SERVICE

One with experience could say:
  1. I have been in little holes-in-the-wall where service was
     quick as a flash and huge dining halls where it took me
     a year to get my soup.

2. The service depends rather on how many waiters there are, how well they are trained, and how quickly they in turn can be served in the kitchen.

Statement 1 is made to show the falsity of the conclusion. The first part of this statement is somewhat beside the point because good service in a small place would not imply poor service in a large one. The second part is more to the point, for it shows that not all large places have quick service. On the other hand, statement 2 shows what is relevant. Both statements depend on experience with restaurants.

Another important factor in any argument is one's degree of self-discipline. If anyone says of our labors, " That was a sloppy piece of work," our immediate inclination is to prove by fair means or foul that he is wrong, one of the most telling "arguments" being that his work was "sloppier." To rule our emotions and to force ourselves to see and take account of relevant but unpleasant factors is no mean achievement. But this again is a matter which we cannot consider here.

Finally, we wish merely to mention the field of mathematical thought which makes it its business to set up a precise language for the classification of logical structure; symbolic logic. A very interesting and useful little book on this subject is Quine's *Elementary Logic* (reference **30** in Bibliography). Here the machinery for examining a logical argument is set up though it is not the author's purpose to show many of its applications. For our purposes in everyday life the old Aristotelian logic is sufficient, and it is this logic with which we are concerned in this chapter. However, that it not always suffices is shown by the following example: suppose a man makes the statement, "I am not telling the truth." There are then two possibilities. First, he is telling the truth, in which case he must be not telling the truth because he said, "I am not telling the truth." Second, he is not telling the truth which is exactly what he said he was doing which implies that he was telling the truth. We should then be forced to conclude that at that time he was neither telling the truth nor not telling the truth.[1]

EXERCISES

Analyze the logical structure of the following fourteen arguments somewhat along the lines of the analysis of the argument at the beginning of this section. Use diagrams where it seems desirable.

1. Bertrand Russell describes the following "reasoning": "Two of my servants were born in March, and it happens that both of them suffer from corns. By the method of simple incomplete enumeration they have decided that all people born in that month have bad feet, and that therefore theirs is a fate against which it is useless to struggle."

2. "Internationalism does not and cannot make for peace because it seeks to keep intact all of those elements which have always made for war. Fortunately that is now becoming apparent. The League of Nations, had it been able to carry out its ludicrous peace policy of enforcing economic sanctions against Italy, would have started a general war in the name of peace!" — Peek, *op. cit.*, p. 348.

3. "As an economic policy, it is impossible, by influencing the extent of planting or breeding, to determine in advance the volume of farm production. As a national policy, it is unwise to attempt to reduce toward the danger point the volume of the production of essential foods and raw materials. The problem, therefore, is control of supply rather than control of production." — *Ibid.*, p. 46.

4. "A man should learn to detect and watch that gleam of light which flashes across his mind from within, more than the lustre of the firmament of bards and sages. Yet he dismisses without notice his thought, because it is his. In every work of genius we recognize our own rejected thoughts: they come back to us with a certain alienated majesty. Great works of art have no more affecting lesson for us than this. They teach us to abide by our spontaneous impression with good-humored inflexibility then most when the whole cry of voices is on the other side. Else, tomorrow a stranger will say with masterly good sense precisely what we have thought and felt all the time, and we shall be forced to take with shame our own opinion from another." — Emerson, essay on "Self-Reliance."

5. "You must love them [the authors of good books], and show your love in these two following ways. 1. First, by a true desire to be taught by them, and to enter into their thoughts. To enter into theirs, observe; not to find your own expressed by them. If the person who wrote the book is not wiser than you, you need not read it; if he be, he will think differently from you in many respects.

[1] See reference **25**, pp. 213 ff.

"Very ready we are to say of a book, 'How good that is — that's exactly what I think.' But the right feeling is, 'How strange that is! I never thought of that before, and yet I see it is true; or if I do not now, I hope I shall, some day.' " — John Ruskin, " Of Kings' Treasuries."

Compare Ruskin's argument with that of the previous exercise. Are they compatible? Give the reasons for your conclusions.

6. A larger proportion of the population of Swarthmore, Pennsylvania, is in *Who's Who* than in any other city of the United States. Hence if you want to get your name in *Who's Who* come to Swarthmore to live.

7. Most people who die, die in bed. Hence, in order to live long, stay out of bed.

8. "Europe has a set of primary interests, which to us have none, or a very remote relation. — Hence she must be engaged in frequent controversies, the causes of which are essentially foreign to our concerns. — Hence, therefore, it must be unwise in us to implicate ourselves, by artificial ties in the ordinary vicissitudes of her politics, or the ordinary combinations and collisions of her friendships, or enmities." — Washington's Farewell Address.

9. "The cruelest lies are often told in silence. A man may have sat in a room for hours and not opened his teeth, and yet come out of that room a disloyal friend or a vile calumniator. And how many loves have perished because, from pride, or spite, or diffidence, or that unmanly shame which withholds a man from daring to betray emotion, a lover, at the critical point of the relation, has hung his head and held his tongue?" — R. L. Stevenson, "Truth of Intercourse" in *Virginibus Puerisque.*

10. "In one half hour I can walk off to some portion of the earth's surface where a man does not stand from one year's end to another, and there, consequently, politics are not, for they are but as the cigar-smoke of a man." — Thoreau, "Walking."

11. "Due to the low temperatures obtained in the chimney when gas heat is applied in highly efficient boilers, there is a possibility that water vapor in the stack gasses will be cooled below its condensation temperature and deposit on the walls of the chimney. The condensed moisture contains small amounts of acid. If the chimney is not tight, the liquid may leak through the chimney walls and discolor the interior walls; it is likely, of course, to corrode the smoke pipe, more so than occurs with other fuels, as a rule. . . . A tile lining inside the chimney will help greatly to prevent this trouble." — *Consumers' Research.*

12. *Portia:* "The quality of mercy is not strain'd.
      It droppeth as the gentle rain from heaven
      Upon the place beneath. It is twice blest:
      It blesseth him that gives and him that takes.

. . . .

It is enthroned in the hearts of kings;
It is an attribute to God himself;
And earthly power doth then show likest God's
When mercy seasons justice. Therefore, Jew,
Though justice be thy plea, consider this,
That in the course of justice, none of us
Should see salvation. We do pray for mercy,
And that same prayer doth teach us all to render
The deeds of mercy."
— *The Merchant of Venice*, IV, i.

13. A kidnapper wrote to a bereft parent: "I will give you back your child if you will tell me truly what will happen to it." What should the parent say? Give your argument.

14. "To have a clear understanding of the war, Americans must see how the fighting goes — the battlefields, soldiers, weapons, geography, people, cities, industries. To have a satisfactory goal, Americans must see what the fighting is *for*, all the manifestations of democracy and our way of life. *Life* shows these things with all the *reality*, *simplicity* and *scope* of life itself — so that every issue is eagerly and easily read by 21,900,000 people." — A *Life* advertisement.

15. In a strange kingdom (very strange it must have been) there was a conviction that all nobles tell the truth and that all slaves lie. Once three men, all dressed alike, appeared before the king. The first one mumbled something. The king said to the second man, "What did he say?" The second said, "He says he's a slave." The king then said to the third, "What is the second man?" The third said, "The second one is a prince." Were there two nobles and one slave or two slaves and one noble? Carefully give your reasons.

16. There was a vacancy in the department of philosophy in a large university. The head of the department brought together three well-known logicians, A, B, and C, who had applied for the position, in order to select one to fill the vacancy. Finding them all qualified, and unable to make a selection, he gave them a final test. He blindfolded them, telling them that he had some soot on a finger. He said, "I will rub a finger over the forehead of each of you. I may leave a smudge on one, two, or three of you, or perhaps on no one. When the blindfolds are removed I want you to give one tap on the floor if you see either one or two smudges. And I want to know which one of you knows that he himself is marked." The blindfolds were applied, the professor left a smudge on each of the three foreheads. Immediately the bandages were removed all three men tapped. After a few moments of silence, Mr. A announced, "I know that I am marked." How did he know? Carefully give your reasons.

17. Solve the following: There are three men, John, Jack, and Joe, each

of whom is engaged in two occupations. Their occupations classify each of them as two of the following: chauffeur, bootlegger, musician, painter, gardener, and barber. From the following facts find in what two occupations each man is engaged:

1. The chauffeur offended the musician by laughing at his long hair.
2. Both the musician and the gardener used to go fishing with John.
3. The painter bought a quart of gin from the bootlegger.
4. The chauffeur courted the painter's sister.
5. Jack owed the gardener $5.
6. Joe beat both Jack and the painter at quoits.

[From Kasner and Newman, *Mathematics and the Imagination*, p. 189. See reference **25** in Bibliography.]

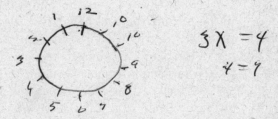

CHAPTER II

The Positive Integers and Zero

---

1. Counting.

A man entering a theater alone says *"one,* please" to the girl selling tickets, and the number 1 has served a useful purpose. Some persons believe that *two* can live more cheaply than *one.* A tripod (if it stands up) has *three* legs. An ordinary city block (of course there are queer blocks in Pittsburgh) has *four* corners, though the same, alas, cannot be said for a city square. Most persons who stay at reasonable distances from buzz saws have *five* fingers on each hand. Every snowflake favors the number *six.* The crapshooter prays for *"seven* come *eleven"* and when we pay for a dozen rolls we expect to have *twelve.* To preserve our luck we refrain from continuing.

The words "one," "two," "three," etc., are, of course, only our names for ideas which are not by any means peculiar to English-speaking peoples. We say "three," the French say "trois," the Germans "drei" — there is a name for it in almost every language. There is even an international sign language for it: the most primitive savage and the headwaiter in a fashionable hotel alike would hold up three fingers and every intelligent being, regardless of race, would understand him; that is, he would comprehend to this extent: whatever the things he was talking about, there would be "just as many" things as there were upraised fingers. The meaning of the phrase "just as many" or its counterpart in another language might be explained in any one of various ways; one could say that it would be possible, given enough string, to connect by strings each thing with an upraised finger in such a manner that each thing would be connected with exactly one finger (but not a specified finger) and each finger would be connected with exactly one thing. One could even dispense with the string and talk abstractly about a "one-to-one correspondence" between the fingers held up and the things. This would mean that by some unspecified method each thing would be associated with or correspond to exactly one finger and each finger to exactly one thing. Similarly, if in an assemblage each man had a hat on, we should know that there were just as many hats as men (no strings are necessary except perhaps to hold the hats on). The fact that we do not then necessarily know either the number of hats or the number of men (though knowledge of one gives knowledge of the other) shows that there is a little more to this process we call **counting** than establishing a one-to-one correspondence. We must have established such a correspondence with some known standard — the notches on a stick, our fingers, or a set of figures learned in a certain order — before we know "how many." Our fingers or a set of numbers merely forms a common yardstick by which we can measure how many things there are in any given set of things. We are particularly fortunate that, when it comes to counting, we need not refer to a standard yardstick in the National Bureau of Standards but carry around with us as a standard a memorized sequence of numbers which can, when the occasion demands it, be reduced to a sign language which all the world understands.

Perhaps the greatest usefulness of counting is the resulting ability to compare the quantity of one thing with the quantity of another thing or perhaps the quantity of one thing at two different times. Given two sets of five people we should know at a glance[1] that the number of people in one set is the same as the number in another set. But probably with even as few as ten people, and certainly with a larger collection it would in general be easier to count each

[1] See the story of the crow, reference **9**. Boldface numbers refer to the Bibliography in the back of the book.

set and compare the numbers (that is, measure the quantities against a common standard) than to try to establish any direct one-to-one correspondence between the members of one set and the members of the other. Certainly the easiest way to compare the number of hens in a coop at one time with the number at another is to count them at the two times.

Possibly the most fundamental property of numbers is that *the number of objects is the same whatever the order in which (or the method by which) the one-to-one correspondence is established.*[1] If the pigeons in a cage were counted by catching each in turn and labeling them with successive numbers, the count would be the same whatever were the order in which they were caught. As a matter of fact, it is precisely this fact which underlies our counting of a set of objects in two different orders (or without regard to the order), as a means of checking one count against the other. It can, of course, be pointed out that when we are counting we do not distinguish between one object and another — they are all alike as far as the enumeration is concerned — and hence the order makes no difference, but that is not really a proof of the fundamental property. To attempt to prove it would then require the proving of something else:

> Great fleas have little fleas upon their backs to bite 'em,
> And little fleas have lesser fleas, and so *ad infinitum*.
> — De Morgan.

One might just as well make the best of the first flea, and not run the risk of calling up all "his sisters and his cousins and his aunts." We might, for instance, be foolish enough to attempt to prove that the sun will rise tomorrow. We could quote evidence to the effect that the sun has risen for the last few days, assert that there have been no signs of the earth slowing down or of the approach of a threatening heavenly body. But by the time we had completed our argument we should have made many assumptions much

[1] See reference **25**, Chap. II.

harder to believe or understand than is the simple assumption that the sun will rise tomorrow. So we accept without attempt at proof the statement that *the number of a set of objects is independent of their order.* It would be a queer world if this were not so.

This property of number is a very unusual one. In most respects the order in which things occur makes a great difference. To exchange the quarterback and center on a football team would usually alter the team considerably though it would still contain eleven men. The batting order of a baseball team is a vital matter. The coat and shirt in either order are *two* articles of clothing but a certain order is to be recommended in putting them on. Persons not under the influence of alcohol usually undress *before* taking a bath. On the other hand, in throwing a seven with two dice, it is of no particular importance which die hits the table (or the ground) first. Nor does it matter in a game of bridge whether the first or the thirteenth card dealt you is your lone ace. Indeed, it is almost paradoxical that, while a number system is an ordered set of marks, its chief use is in counting which, in a large degree, is itself independent of order.

These numbers which we get by counting things and write as 1, 2, 3, 4, ⋯, are called **positive integers** (they are sometimes called **natural numbers**). (Three dots means "and so forth" — a dot for each word.) The necessity for such a name instead of the mere appellation "number" is that later on we shall want to designate as numbers ideas which cannot be obtained by counting things, though they share many properties with the positive integers. Why we call them "positive integers" instead of "glumpers" or any other name can best be explained when we come to discuss other kinds of numbers.

Indeed, there is one new "number" not obtainable by counting, which it is convenient to introduce immediately: the number **zero,** whose symbol is 0. In our younger days we quite often read 0 as "nothing," and in many circumstances that is quite a proper meaning. (Reading it "naught" is merely an old-fashioned way of saying the same thing.) As a matter of fact this is the only sense in which, for a long time, the number 0 was used. Just as I can tell you how many troubles I have, if I have some, by mentioning a number which is a positive integer, I find it convenient to use the number zero if I have none. As far as counting goes, zero means "none" and though we shall find later that this idea of zero is not adequate in some situations, we shall climb that hill when we come to it. One cannot, strictly speaking, get zero by counting things and hence it is not a positive integer; but it is a number with many of the same properties, as we shall see. In place of the long phrase "positive integer or zero" we shall often use the word "number" in this chapter; but we shall be careful to make no statements about "numbers" which are not true for all numbers we consider in this book even though our proofs and discussions are solely in terms of positive integers and zero.[1]

EXERCISES

1. In which of the following does order play an important part, and in which is it a matter of indifference? If in any case the answer is "it depends," argue the pros and cons: the courses of a meal; the drum major and his band; the items of a shopping list; the names in a telephone directory; marks in the six tests of a course; painting the various parts of a floor; writing to one's three sweethearts; the hen and the egg; putting gasoline and water into a car; mixing the ingredients of a cake.

2. Give two not too obvious examples in which order does and two in which it does not make a difference.

3. The word "pair" is not applied to *any* two objects but to two which are more or less alike — a cow and a horse would not constitute a pair nor would even a roan horse and a bay horse. The term "sextet" is generally associated with music. Give five other examples of such restricted number names.

4. It is stated above that 0 does not always mean "nothing." Give three examples to support this statement.

5. If the number of people in one set is less than the number in another, what precisely does this mean in terms of a one-to-one correspondence?

[1] See topics 1 and 2 at the close of the chapter.

6. Using the results of exercise 5, show that if the number of people in set A is less than the number in set B, and if the number in B is less than the number in set C, then the number in set A is less than the number in set C.

7. What kind of correspondence describes the phrase "twice as many" in a manner analogous to the way in which one-to-one correspondence describes "just as many"? State precisely what you mean by this new kind of correspondence and give two examples of it.

2. The commutative property of addition and multiplication.[1]

One of the first things we found after learning to count, and probably knew in a practical way before we could count very far, was that if you have one apple and I have two, then we both together have three; and the same is true if we have donkeys or elephants. That was the beginning of that onerous but so useful process of addition. One of its properties that came into our consciousness later but probably without much effort was the fact that it makes no difference in which order the addition is performed (order sticking its head in again). One and two have the same sum as two and one. This fact follows immediately from the fundamental property mentioned in the last section. If we let dots represent the objects as follows:

●   ● ●

since the order of counting makes no difference, the result of counting from the left (one and two) is the same as that of counting from the right (two and one). There is, to be sure, something a little new in grouping the objects, but fundamentally it is this old question of order again. The same device can be used for any two sets of objects to establish the property that *the order of addition of two numbers makes no difference to the sum.* We express this briefly by saying: **addition is commutative.** In symbols this amounts to writing

$$a + b = b + a$$

when a and b are any positive integers. This law or property holds even if one or both of the numbers of the sum is zero;

[1] Cf. reference **31**, Chap. III.

since, for example, if you have three dogs and I have none, we both have three and $3 + 0 = 3$; and the total number of dogs is the same if I have three dogs and you have none. Also, if each has none we both together have none.

That multiplication is commutative is not nearly so apparent. At the outset, notice that if each of three persons has five peanuts we could find by counting that altogether they had fifteen peanuts. When the three pool their resources, the five peanuts are multiplied by three. This is a concrete instance of the abstract statement that three times five, or five *multiplied by* three, is fifteen. We write it $3 \cdot 5 = 15$. (In this book we indicate multiplication by a dot as in $4 \cdot 5 = 20$. In the case of the product of two letters or a numeral and a letter we usually omit the dot and write ab instead of $a \cdot b$ and $5\,b$ instead of $5 \cdot b$. Notice that the dot for multiplication is differently placed from the period.) When we learned the multiplication table we merely memorized numerous results of such multiplications so that once we knew that each of three persons had five peanuts we deduced immediately the fact that there were fifteen peanuts in all; counting in this case became unnecessary. We also learned that five times three is fifteen. But, since five men with three peanuts each look different from three men with five peanuts each, it was probably some time before we realized that it was any more than a coincidence that the total number of peanuts in one case was the same as that in the other. We begin to see that this is indeed more than a coincidence when we consider the following diagram.

There are three rows of five dots each — three times five; looking at it another way, there are five columns of three each — five times three. The number of dots is the same whichever way you look at it. We could do the same thing if there were 20 rows of 16 dots each. In fact, if we let

a stand for some positive integer and b for some other (or the same) positive integer, a rows of b dots each present the same picture as b columns of a dots each; in other words, a times b is equal to b times a no matter what positive integers a and b are. Hence we say that **multiplication of positive integers is commutative.** Briefly, we may then write

$$ab = ba$$

where ab is understood to mean a times b and ba is b times a. The fact that multiplication by zero is also commutative will appear below.

It should be seen that our diagram of dots does something which could be accomplished by no amount of comparing the product of a pair of numbers in one order with that in the other order. No matter how many examples one gives to show that one number times another number is equal to the other times the one, one could not be sure but that some untried pair might fail to have this property. But we can establish the conclusion for *every* pair of numbers by seeing that *no matter what pair* we have, we can, in imagination at least, make a diagram of dots to fit these numbers and by counting the dots in two different orders establish our result. A similar situation outside of mathematics is the following: suppose I wished to prove that I can *always* distinguish between alcohol and water. You might try my knowledge on samples of one or the other until doomsday without *proving* that I could distinguish between them though every successful example would lead you nearer to admitting my powers of discrimination. But if I told you that my test would be trying to light the liquid with a match, you would then know that *whatever* sample was presented to me, I could tell whether it was alcohol or water; furthermore, this could be established without any trial whatsoever. Thus, while we often use examples to lead one to suspect an underlying reason, it is the latter and not the examples which proves that something *always* holds.

It is interesting to notice that the number 1 plays the

same role for multiplication that the number 0 plays for addition, for, no matter what number b is,

$$b + 0 = b \quad \text{and} \quad b \cdot 1 = b$$

are both true.

Now is a good time to consider multiplication by zero. It is at this point that the idea of zero meaning "nothing" begins to be inadequate. If 20 people have 3 apples apiece, all together they have $20 \cdot 3 = 60$ apples. If 20 people have no apples apiece, we should say without hesitation that all together they would have no apples and hence $20 \cdot 0$ should be 0. But we begin to have difficulty if we turn this multiplication around and ask: if no people have 20 apples apiece how many do they have altogether? The question immediately arises: "What do you mean by 'no people having 20 apples apiece'?" If there are no people, there is no sense in talking about how many apples they have. Now this perplexity is really an advantage because whatever we define this statement to be we have no apparent conflict with our previous ideas about the nothingness of zero. It is good to have multiplication by zero commutative and so we define $0 \cdot 20$ to be 0. And for every positive integer b we agree that $0 \cdot b = b \cdot 0 = 0$. Furthermore, we do not like exceptions and also agree that $0 \cdot 0 = 0$. Hence for every number b in this chapter we agree that

$$0 \cdot b = b \cdot 0 = 0.$$

EXERCISES

1. Which of the following are commutative? If the answer is "it depends," argue the pros and cons: taking out insurance and having an accident, shooting the right and left barrels of a shotgun, the cart and the horse, the locking of the barn and the coming of a thief, dusting the furniture and mopping the floor, clipping the inside or outside of a hedge, putting a stamp on an envelope and sealing it. Give reasons.

2. Form an addition table of all pairs of numbers from 1 to 10 making full use of the fact that addition is commutative.

3. Why is the following statement true: knowledge of the fact that multiplication is commutative makes almost half of the multiplication table unnecessary.

4. Show that if a and b are positive integers, then ab and $a + b$ are positive integers and hence are not zero.

5. Suppose each of a and b is zero or a positive integer. If $ab = 0$ what can you conclude about a and b? What if $a + b = 0$?

3. The associative property of addition and multiplication.

As soon as we begin to count objects in more than two sets we have another property which again has something to do with order but in a different way from the commutative property considered above. If we have dots in the following sets

•   • •   • • • •

we can count them first by associating the first two sets whose combined number is three and, considering them as one set of three, add this set of three to the remaining one of four and get the total number seven. If, on the other hand, we had first associated the last two sets and considered them as one set of six dots; then, adding this number to the number in the first set we have the sum seven — the same number as before. This is not particularly surprising but it is an important property or law of addition. So, we say, in brief, that **addition is associative**. This property may be expressed as follows:

$$(a + b) + c = a + (b + c)$$

where the left side means: add a and b and then add c to the result, while the right side means: add b and c and then add this sum to a. The parentheses are thus used to indicate what is done first. Notice that the order in which the letters appear is the same on both sides of the equation, the sole difference lying in the placing of the parentheses.

We should digress to explain at this point that parentheses work from the inside out like the ripples from a pebble thrown into a quiet millpond. For instance,

$$[(a + b) + c] + d$$

means: add a and b, add this result to c, add that result to d. Here, due to the associative law, it makes no difference which is done first but we shall soon find cases in which it does make a difference. We omit the dot for multiplication when a number is multiplied by a quantity in parentheses or when two quantities in parentheses are multiplied. For instance, we write $3(4 + 5)$ instead of $3 \cdot (4 + 5)$ and $(7 \cdot 8)(3 + 4)$ instead of $(7 \cdot 8) \cdot (3 + 4)$.

To return to the associative property of addition, notice that this is the property which enables us to find by addition the total number of dots in four sets for it reduces the work to computing three sums of two numbers each. It is probably true that we accept this property a little more easily than the commutative property — perhaps because we do not come upon it until a later stage in our development. Nevertheless, associativity is not by any means a universal property. Thus, in making iced tea we "add together" tea leaves, hot water, and ice, but (tea leaves) + (hot water) = tea and (tea) + (ice) = (iced tea) while (hot water) + (ice) = tepid water and (tea leaves) + (tepid water) is not very good. But in addition, altering the order of association of the parts does not change the results. Fundamentally we can never add more than two numbers at a time, though, with practice, we can sometimes almost immediately get the sum of three or more. This property which we have just illustrated gives sense to the question: what is one and two and four? While the question may mean one of two things (is it one added to the sum of two and four, or the sum of one and two added to four?) there is no ambiguity in the result. Notice that, in contrast to commutativity which has to do with interchanging the order of addition of *two* numbers, associativity has to do with selecting a pair from a set of *three or more* numbers.

Multiplication of positive integers is also associative. Imagine three trays of glasses and suppose in each tray the glasses are arranged in the following pattern:

Consider the trays piled vertically. To find the total number of glasses we can count those in one tray by multiplying 4 by 5, getting 20; thus we have 20 stacks of glasses with 3 in each stack. This method of getting the total number could be indicated by $(4 \cdot 5)3 = 60$; the parentheses indicating that it is the product of 4 and 5 which is multiplied by 3, that is, we multiply 4 by 5 and then multiply the result by 3. There is another way in which we could count the glasses. Looking at the pile from the front we should see 5 columns of 3 glasses each — hence 15 glasses. From our knowledge of the piles we should know that there were 4 such "slices"; hence 4 fifteens. This method could be indicated by $4(5 \cdot 3) = 60$. Again we are merely associating numbers in a different way. It is interesting to notice, by the way, that we can write the numbers in the same order but arrange the parentheses in a different way. Looking at the pile from the side would give us a combination of the commutative and associative property; namely, $(4 \cdot 3)5 = 60$.

Briefly, the associative property of multiplication states that

$$(ab)c = a(bc).$$

That this product is also equal to $(ac)b$ results from the associative and commutative properties. This property of associativity does for multiplication just what it did for addition; namely, it allows us to multiply as many numbers as we please by finding the product of any two, then the product of this result with another, and so on. Hence there is no ambiguity so far as the result is concerned in writing $4 \cdot 5 \cdot 3 = 60$ without specifying which pair is multiplied first. Therefore in the product of three or more numbers we need no parentheses.

It is sometimes convenient to use braces { } and brackets [ ] for, though there is no gain in clarity in the expression $(5 + 6)[7 + 8]$ over $(5 + 6)(7 + 8)$, some confusion is likely to result when parentheses occur within parentheses. For instance, in the expression

$$N = 4((3 \cdot (5 + 6) + 7) \cdot 2 + 6)$$

it is a little awkward to pair the parentheses correctly though, if one works from the inside out, there is no real ambiguity. It is, however, clearer to write it in the form

$$N = 4 \cdot \{[3 \cdot (5 + 6) + 7] \cdot 2 + 6\}.$$

We then work from the inside out as follows

$$
\begin{aligned}
N &= 4\{[3 \cdot 11 + 7] \cdot 2 + 6\} \\
  &= 4\{40 \cdot 2 + 6\} \\
  &= 4 \cdot 86 \\
  &= 344.
\end{aligned}
$$

EXERCISES

1. Find the value of $[(2 \cdot 3)(4 + 5)] + 6$.   *Ans.* 60.

2. Find the value of
   a. $\{(2 + 5)(6 \cdot 2)\} + 1$.
   b. $[3(2 + 5)] + [(3 + 6) + 7]$.
   c. $\{(2 + 6)(7 + 5)\} + 3$.

3. Show that $5(7 + 6) = (5 \cdot 7) + (5 \cdot 6)$.

4. Write the following as an equation using parentheses: a times the sum of b and c is equal to the sum of a times b and a times c.

5. Express in words the following:
   a. $[(a + b) + c] + e$.
   b. $[c(ab)]d$.
   c. $[(xy) + z]t$.

6. What parentheses in exercises 2 and 5 above does the associative property of addition and multiplication render unnecessary?

4. The distributive property.

So far, the properties which we have considered have applied equally well to addition or to multiplication but to only one at a time. One must use care in combining addition and multiplication. For instance, $(4 \cdot 5) + 3$ is not equal to $4(5 + 3)$ as can be seen by performing the two operations. One cannot, then, combine multiplication and addition in a kind of associative law. But there is a property of numbers which involves both addition and multiplication. If we add 2 and 3 and multiply the sum by 6, we get the same result as if we had multiplied 2 by 6 and multiplied

3 by 6 and added the two products. We can write this as follows:

(1) $(2 + 3)6 = (2 \cdot 6) + (3 \cdot 6).$

This can be shown diagrammatically, as were the other properties, for any three numbers, and this is left to the reader. Such a diagram is important in seeing that such a process works for any three numbers — this is very different from finding for any like expression that the result of the work on one side of the equality for the particular numbers used is the same as the result of the work on the other side. This property is described by saying that **multiplication of numbers is distributive with respect to addition,** and it means that for any three numbers a, b, and c

(2) $a(b + c) = (ab) + (ac).$

The reason for the terminology of the boldface phrase is that there is a kind of distribution of the multiplication throughout the addition. This may seem to be a rather awkward way of expressing the property but it is doubtful whether there is any really shorter way of accurately describing it. Even though we briefly refer to this as the **distributive property** the longer designation is important for it emphasizes the relative roles of addition and multiplication. If addition were distributive with respect to multiplication we could interchange $+$ and $\cdot$ in equation (1), which would give us

$$(2 \cdot 3) + 6 = (2 + 6)(3 + 6)$$

that is,

$$12 = 72,$$

which is false. In fact, *addition is usually not distributive with respect to multiplication.*

Here is an example outside of mathematics for which the distributive property does not hold: suppose we let "touching" play the role of multiplication and "combining chemically" the role of addition. Corresponding to the left-hand side of equation (2) we might then have touching a lighted match to a combination of hydrogen and oxygen, that is, to water; corresponding to the right-hand side we should have touching a lighted match to hydrogen, touching it to oxygen, and combining chemically the results. The end result in the two cases would be very different.

This property is the hardest of the three to comprehend since it does not appear on the surface of our everyday operations with numbers. That it really underlies our mechanical processes can be seen by examining our process of multiplying 536 by 7. Dissected, it might look like this with the reasons indicated briefly in the right-hand column.

(i) $536 \cdot 7 = (500 + 30 + 6)7$		Decimal notation
(ii) $= (500 \cdot 7) + (30 \cdot 7) + (6 \cdot 7)$		Distributive property
(iii) $= (500 \cdot 7) + (30 \cdot 7) + (40 + 2)$		Multiplication Decimal notation
(iv) $= (500 \cdot 7) + [(30 \cdot 7) + 40] + 2$		Addition associative
(v) $= 3500 + 250 + 2$		Multiplication and addition
(vi) $= 3500 + 200 + 52$		Addition associative
(vii) $= 3700 + 52$		Addition associative
(viii) $= 3752$		Addition

Notice that the "carrying" part of the process consists in getting equation (v) and (vii) above. Strictly speaking, the associative property of addition enters in all but the last step above since otherwise the sums would be ambiguous.

At this point let us introduce a little convention which saves writing. To make sure that there is no ambiguity we have written $(3 \cdot 5) + 2$ to certify that 3 is multiplied by 5 and the result added to 2. This has, of course, an entirely different result from that of $3(5 + 2)$ in which we add 5 to 2 first and then multiply by 3. By agreeing always to give multiplication precedence over addition — that is, always multiplying before adding whenever there is no indication to the contrary — we can write $3 \cdot 5 + 2$ to mean $(3 \cdot 5) + 2$. *It is now all the more important that we retain the parentheses when we mean that the addition should be made first.* Using this convention would enable us to omit all the parentheses in lines (ii) to (viii) above but not in line (i).

The distributive property is the source of much of the difficulty in algebra. On this account it is especially important that we establish it firmly in terms of numbers. Notice that it may now be written

$$a(b + c) = ab + ac.$$

EXERCISES

1. Use parentheses to express the associative property of addition for a particular numerical example.

2. In adding a sum of ten numbers how many sums of two numbers would it be necessary to find? What would be your answer for a corresponding problem in multiplication?

3. If each of a, b, c is a positive integer or zero, show that $(ab)c = a(bc)$, assuming the equality for positive integers.

4. Remove all the parentheses you can from the following expressions without making them ambiguous, and find their values:
a. $(3 \cdot 5) + (4 + 5) + 3(4 + 5) + (4 + 5)(3 + 4) + (3 \cdot 5)6 + (3 \cdot 5)6 + (3 \cdot 5)(6 + 2)$.
b. $(2 \cdot 3) + 2(1 + 7) + 3 + 7 \cdot 6 + 4(3 + 2) + (7 + 5)(6 + 3)$.

5. In the expressions of exercise 4, replace every multiplication by addition and every addition by multiplication. Then do for these expressions what was required in the previous exercise.

6. In the following expression insert one pair of parentheses in as many ways as possible (there are 15 ways) and evaluate the resulting expressions:

$$1 \cdot 2 + 3 \cdot 4 \cdot 5.$$

How many different results are there? Which are equal to the expression as written without parentheses?

7. Do Exercise 6 with addition and multiplication interchanged.

8. Do Exercise 6 inserting two pairs of parentheses instead of one.

9. In the equation $a(b + c) = ab + ac$ just preceding the exercises, could the parentheses be omitted? Explain the reasons for your answer.

10. Dissect the process of multiplying 827 by 3, giving the reasons for each step and including only necessary parentheses, as was done on page 40.

11. Show that $(2 + 4)(3 + 5) = 2 \cdot 3 + 4 \cdot 3 + 2 \cdot 5 + 4 \cdot 5$. What will be the corresponding expression for $(a + b)(x + y)$?

12. Show that $5 \cdot 3 + 5 \cdot 2 + 5 \cdot 6 = 5(3 + 2 + 6)$ and find the corresponding right-hand side of the following equation:

$$ax + ay + az =$$

13. Pick out all correct right-hand sides for each of the following equations and give reasons for your answer:

$$(1) \qquad ax + ay + bx + by = \begin{cases} (a + b)(x + y) \\ a(x + y) + (x + y)b \\ a(y + b) + x(a + y) \end{cases}$$

$$(2) \qquad ar + 2\,a = \begin{cases} r(a + 2) \\ a(r + 2) \\ a(r + 1) + a \end{cases}$$

$$(3) \qquad ab + ac + bc = \begin{cases} a(b + c) + bc \\ b(a + c) + ac \\ a(b + c + a) \end{cases}$$

$$(4) \qquad P + Pi = \begin{cases} P(1 + i) \\ i(1 + P) \end{cases}$$

$$(5) \qquad P(1 + i)^2 + P(1 + i)^2 i = \begin{cases} P(1 + i)^3 \\ P(1 + i)(1 + i^2) \\ Pi + P(1 + i)^3 \end{cases}$$

14. $(3 + 4)8 = 8 \cdot 3 + 8 \cdot 4$. That a like expression would hold for any three numbers would follow from what properties of numbers?

15. Is $1 \cdot 5 + 3 = 1(5 + 3)$? Will this still hold if the 1 is replaced by a 4? For what positive integers a, b, c is $ab + c = a(b + c)$?

16. Use dots to prove the distributive law.

* 17.[1] Can you find three positive integers or zero for which addition is distributive with respect to multiplication? If not, can you prove that such numbers do not exist?

5. The number system to the base ten.

"A rose by any other name would smell as sweet" but not nearly so many people would know what you were talking about if you called it a *Rosa spinosissima*, and to some sensitive souls, a euphonious name even contributes something to the smell. Many will admit that a really good name is always partially descriptive. No matter what you call the numbers, they have the properties we have talked about but it would be incongruous, not to mention other objections, if the names of numbers — the essence of order — were themselves without reason (rhyme is not necessary). One could devise one hundred random names for the numbers from 1 to 100 but it would certainly retard the education of our youth; worse, our conception of how many a hundred men are would be very vague. We can visualize twenty men fairly well and to say that there are five sets

---

[1] Starred problems are within the range of only the best students.

of twenty men each, whether or not they are actually bunched in that way, gives a pretty good mental picture of how many a hundred men are. Furthermore, such a grouping would make it much easier to count them. An efficient bank teller does not count your handful of pennies one by one but in sets of five. He thereby not only works more quickly but he is surer of his result. There is a disadvantage in having too small a set if one has a large number to count, because then the number of sets mounts. On the other hand, it is awkward to have the set too large. Grouping by tens is natural to us who can wiggle our fingers. The fact that we count by tens and not by twenties may be an argument for the anti-evolutionists, or it may merely indicate that when an ape had more than ten children she could not be bothered keeping track of them all. However, the appearance of the word "score" is damning and the French come in for their share of suspicion with their *vingt* and *quatre-vingt*. At any rate we can be proud of our ancestors in that they hit on the best way to extend the notion of number by talking of tens of tens and later of thousands of thousands.

Though we think of the method of counting by tens in terms of the numerals we use, there were many other systems of numeration which were used to count by tens. The Romans and others before them counted by tens. It is usually pointed out that their numerals are awkward because it is too difficult to calculate with them, but the Romans probably did not use them for that purpose except for the simplest calculations. For such purposes they used the abacus, that harp of the merchant, the origins of which go back to the fifth and sixth centuries B.C. However, the Hindu-Arabic numerals, which most of the civilized world uses today, have two real advantages over the Roman system. The first is the greater compactness of the symbols for the numbers from 1 to 9. The second, namely, the giving of place value, would be applicable in the case of any nine symbols with zero and came relatively late in the develop-

ment of our system of notation. In many of the early notations there was no relation between the symbol for the
number 7 and that for 70. The Greeks and the Jews wrote
for a time 700.80.9 for the number which we should write
as 789. We are so accustomed to this notation of ours that
it requires sometimes a little effort to recall that the right-
hand **digit,**[1] 9, gives the number of units, the one to the
left of it, 8, the number of tens, and the one to the left of it
the number of hundreds. The symbol 789 is a very elegant
way of writing 700 + 80 + 9. In our notation the use of
zero plays an important part in denoting the absence of
tens, for instance, in the number 304. This use of zero
seems to have been very hard for the human race to in-
vent.

Perhaps we shall appreciate the advantages of our notation more fully if we dissect this process of addition to see
how it is made. To find the sum of 12 + 34 + 56 we write
it 10 + 2 + 30 + 4 + 50 + 6 using the associative property of addition. Since addition is associative and commutative, we can rearrange the terms to have 10 + 30 + 50 +
2 + 4 + 6. The sum of the last three is 12 or 10 + 2 and
our sum becomes 10 + 30 + 50 + 10 + 2. Adding the first
four numbers we have 1 + 3 + 5 + 1 tens, which is 10 tens,
or 100. Adding this to 2 we have our result, 102. This
whole process is, of course, facilitated by writing the numbers in a column

$$12$$
$$34$$
$$56$$

By reading down (or up) the right-hand column and then
down the left-hand column we have the rearrangement necessary to adding the tens and the units. We add the right
column first because it may contribute to the left column.
"Carrying the one" is acknowledging this contribution.
This is perhaps more clearly brought out by writing the sum

---

[1] We say, for example, the number 3572 has the "digits" 3, 5, 7, and 2.

in the form which one often uses when adding a long column of figures in order to make checking easier:

$$
\begin{array}{r}
12 \\
34 \\
56 \\
\hline
12 \\
9 \\
\hline
102
\end{array}
$$

Here the sum of the right-hand column appears explicitly. The 9, of course, is really a 90, but its position makes the writing of the 0 unnecessary.

To find the product of 4 and 126, we have $4(100 + 20 + 6)$ which, by the distributive property, is $4 \cdot 100 + 4 \cdot 20 + 4 \cdot 6 = 400 + 80 + 24$ which, by the process of addition, is 504. The connection between this and the usual process can be emphasized by writing it in the following manner:

$$
\begin{array}{r}
126 \\
4 \\
\hline
24 \\
8 \\
4 \\
\hline
504
\end{array}
$$

Zeros are not necessary to indicate the positions of 8 and 4 in the fourth and fifth lines of the scheme and hence are there omitted. The process of "carrying the 2" and then the 1 is, of course, a means of using one's head instead of the paper and pencil.

6. The number system to the base twelve.[1]

Without seeking to change the habit of counting by tens, that is, the use of the **decimal system** (a **number system to the base ten,** as it is sometimes called), we can consider the advantages, from certain points of view, in other systems. It is convenient to have twelve inches in a foot, for then a

[1] See topic 3 at the close of the chapter.

quarter of a foot is three inches. This is much more comforting than the situation in the metric system of measurement for a quarter of a decimeter is two and one-half centimeters and we like to avoid fractions. For a similar reason, dividing the hours into sixty minutes is very helpful, for though a division into one hundred minutes would give a whole number of minutes for a quarter of an hour, it would not for a third. With sixty minutes in an hour, there is a whole number of minutes in each of the following parts: one-half, one-third, one-fourth, one-fifth, and one-sixth; and who wants to divide an hour into seven parts anyway? There are dozens of times when we find counting by twelves useful. The horses in *Gulliver's Travels*, having three toes on a foot, would perhaps have counted by twelves.

What alterations would be necessary if we had twelve fingers instead of ten? To begin with, we could manufacture an entirely new set of symbols but that would require much mental labor in devising them and in memorizing the results as well as being a trial to the printer. And, after all, the numbers from one to nine can just as well be represented by the same symbols as before. However, if we are to take advantage of place value and indicate our twelves and twelves of twelves by the position of the symbols, we do need new symbols for the numbers ten and eleven. Let us call them t and e. Then our numbers would be:

Base 10: 1, 2, ···, 8, 9, 10, 11, 12, 13, 14, ···, 21, 22, 23, 24, 25, ···
Base 12: 1, 2, ···, 8, 9, t, e, 10, 11, 12, ···, 19, 1t, 1e, 20, 21, ···

where now 10 stands for twelve, 11 for thirteen, 20 for twenty-four (being two twelves) and so forth. We could call this a **dozal** (pronounced "duzzal") system[1] instead of a decimal system since in it we count by dozens instead of tens and, to avoid confusion between the two systems and to emphasize the connection of the names with the notation, let us call the above list of numbers to the base twelve:

[1] A more erudite term is "the duodecimal system."

one, two, ···, eight, nine, ten, eleven, dozen, doza-one, doza-two, ···, doza-nine, doza-ten, doza-eleven, two dozen, two doza-one, ···.

In the dozal system 100 is not one hundred but a *gross*, and 1000 a *great gross*. We shall not attempt to give names further.

Regarding the dozal system as a new language it is apparent that we need some means of translating into and from this system. We illustrate the method by the following:

Example 1. How is the number 7te in the dozal system written to the base ten?

SOLUTION: 7te means: seven gross, ten doza-eleven. This to the base ten is

$$7 \cdot 144 = 1008$$
$$10 \cdot 12 = 120$$
$$11 = \underline{11}$$
$$1139$$

Example 2. The number which is written 1437 in the number system to the base ten is what in the dozal system?

SOLUTION: Just as 1437 means $1 \cdot 10^3 + 4 \cdot 10^2 + 3 \cdot 10 + 7$ (where 10^3 means $10 \cdot 10 \cdot 10$, etc.) when written to the base ten, so if we write it in the form

$$a \cdot 12^3 + b \cdot 12^2 + c \cdot 12 + d,$$

it becomes *abcd* to the base twelve where *a*, *b*, *c*, and *d* are properly chosen numbers from 0, 1, 2, ···, 9, t, e. Now $12^3 = 1728$ which is greater than 1437 and which shows that $a = 0$. The greatest multiple of 144 less than 1437 is $1296 = 9 \cdot 144$. Hence $b = 9$. We then have $1437 = 9 \cdot 144 + c \cdot 12 + d$. Hence *c* is the number e, *d* is 9, and the given number is 9e9 when written to the base twelve. This process may be abbreviated as follows:

$$
\begin{array}{ll}
1437 & \\
\underline{1296} & 9 \\
141 & \\
\underline{132} & e \\
9 & 9
\end{array}
$$

There are two methods of addition in the dozal system which are about equal in difficulty. To add 9 and 8 first we could say, "9 plus 8 is seventeen which is 5 more than twelve and hence is doza-five"; second, we could say, "9 lacks 3 of being a dozen and 8 minus 3 is five which implies that the sum is doza-five." On the other hand, in multiplication we must either learn a new multiplication table or translate back and forth each time between the two number systems. Unless one is going to perform many calculations in the dozal system, the latter is the easier alternative. Then, for example, 9 times 8 is seventy-two which is six dozen.

Let us see how addition in the dozal system goes.

$$\begin{array}{r} t7 \\ 6e \\ \underline{et} \end{array}$$

Seven and eleven are doza-six and ten is two-doza-four (ten being six and four). At the foot of the right-hand column put a 4. Carry two and ten is dozen and six is doza-six and eleven is two doza-five. Hence the sum is 254: two gross, five doza-four. To check our result we could translate the numbers to the decimal system and add as follows:

$$\begin{array}{r} 127 \\ 83 \\ \underline{142} \\ 352 \end{array}$$

$352 = 2 \cdot 144 + 5 \cdot 12 + 4$ which checks with our result in the dozal system. One easily discovers certain aids analogous to aids in the decimal system. For instance, to add eleven to any number increases the dozas by one and decreases the units by one; also it is convenient to group numbers whose sum is dozen whenever possible.

Multiplication is a little more difficult. For example,

$$\begin{array}{r} 5e8 \\ \underline{7} \\ 35^{6}9^{4}8 \end{array}$$

Seven times eight is four doza-eight, seven times eleven plus four is six doza-nine, seven times five plus the six (carried) is three doza-five in the first two places.

EXERCISES

1. Write out the multiplication of $234 \cdot 56$ in such a way as to emphasize the properties of numbers used in deriving the result.

2. Are the properties or laws of addition and multiplication mentioned in sections 2, 3, and 4 the same for the dozal system as for the number system to the base ten? Give reasons.

3. The following are numbers written in the dozal system. Write their expressions to the base ten and check your results.

$\quad\quad$ *a.* eee. $\quad\quad$ *b.* t5e. $\quad\quad$ *c.* tete.

4. The following are numbers written to the base ten. Write them to the base twelve and check your results.

$\quad\quad$ *a.* 327. $\quad\quad$ *b.* 4576. $\quad\quad$ *c.* 17560.

5. Each of the numbers in the following sums are written in the dozal system. Find the value of each sum in this system and check by translating the given numbers and the answer into the decimal system.

a. t6e	*b.* 7te	*c.* 95e
420	80	671
6e1	t19	eee

6. Find the following products of numbers written in the dozal system. Check as in the previous exercise.

$\quad\quad$ *a.* $3t4 \cdot 23$. $\quad\quad$ *b.* $eee \cdot te$. $\quad\quad$ *c.* $6et \cdot 51t$.

7. What is the greatest integer in the dozal system less than a great gross? Find its value in the decimal system.

* 8. In the decimal system, every number which is a perfect square and whose last digit is 0 has 00 as its last two digits. Find a number in the dozal system which is a perfect square, whose last digit is 0 but yet whose next to the last digit is not 0.

7. Number systems to various bases.[1]

The number system in which addition and multiplication are easiest is the **binary** or **dyadic system,** or number system to the base 2. The multiplication table is

$$0 \cdot 0 = 0 \cdot 1 = 0, \quad\quad 1 \cdot 1 = 1$$

and the addition table is

$$0 + 0 = 0, \quad 0 + 1 = 1, \quad \text{and} \quad 1 + 1 = 10.$$

[1] See topic 3 at the close of the chapter.

Since
$$19 = 1 \cdot 2^4 + 0 \cdot 2^3 + 0 \cdot 2^2 + 1 \cdot 2 + 1$$
this number to the base 2 is written: 10011. Similarly 9 is 1001. Adding, we have

$$
\begin{array}{r}
10011 \\
1001 \\
\hline
11100
\end{array}
$$

which we may check by translating back to the decimal system. Then our number becomes
$$0 + 0 \cdot 2 + 1 \cdot 2^2 + 1 \cdot 2^3 + 1 \cdot 2^4 = 28.$$
Multiplication is equally simple:

$$
\begin{array}{r}
10011 \\
1001 \\
\hline
10011 \\
10011 \\
\hline
10101011
\end{array}
$$

which may be checked by translation. We, of course, pay for all this ease by increasing the writing involved in exhibiting the numbers. The seemingly prodigious result of our multiplication is only 171 when written to the base ten.

At the other extreme, one million is written 10 in the number system to the base one thousand but a multiplication table with $999 \cdot 999$ entries is enough to make the stoutest hearts quail. Our number system is a reasonable compromise: the multiplication table is not too long and small numbers do not take too much writing.

The number system to the base 2 enters very usefully into numerous games and tricks. In the next section we shall discuss in detail one such game. A rather interesting trick involving the binary system is the following.[1] Construct four cards like this:

1	2	4	8
1 3 5 7	2 3 6 7	4 5 6 7	8 9 10 11
9 11 13 15	10 11 14 15	12 13 14 15	12 13 14 15

[1] Cf. topic 4 at the close of the chapter.

You as the "magician" ask a friend to think of some number from 1 to 15 and pick out the cards on which it occurs. You then tell him what number he is thinking of. You perform this miracle merely by adding the numbers at the top of the chosen cards. For instance, if he presents you with the cards headed 1, 2, and 4 the number which he was thinking of must have been $7 = 1 + 2 + 4$. The secret of the construction of this table can be seen by writing the numbers 1 to 15 in the binary system.

EXERCISES

1. Write 345 in the number system to the base 2, to the base 5, to the base 15.   *Ans.* 101011001; 2340; 180.

2. Write 627 in the number systems to the bases 2, 5, and 15.

3. Extend the trick given in the last section to five cards and the numbers 1 to 31.

4. What is the connection between the choice of numbers on the cards and writing them to the base 2?

5. What in the number system to the base 7 would be the number 51 in the decimal system? What would it be in a system to the base 2?

6. Write the numbers 36 and 128 of the decimal system in the number system to the base 2 and check your result by expressing the sum in the decimal system. Do the same for the product.

7. Give a test for divisibility by 8 for a number in the binary system.

8. The numbers 55 and 26 are written in the number system to the base 7. Find their squares in the same system.

9. What would be some advantages and some disadvantages in a number system to the base 8?

10. What in the decimal system is the number 4210 as written in the number system to the base 5?

11. Devise a trick analogous to that using the four cards above, where now the number system to the base 3 is involved instead of the binary system and the cards serve to select some number from 1 to 26.

12. Another interesting application of the number system to the base 2 is contained in the following device which reduces all multiplication to multiplication and division by 2. It is a method used by some Russian peasants in very recent times. For example, to find $37 \cdot 13$, write:

$$
\begin{array}{cl}
13 & 37 \\
6 & 74 \times \\
3 & 148 \\
1 & 296 \\
\hline
& 481
\end{array}
$$

The left column is obtained by dividing by 2, disregarding the remainder, and the right by multiplying by 2. In forming the sum, 74 is omitted because it occurs opposite the even number 6. Why will this process always work?

13. Determine a set of six weights which make possible weighing loads from 1 to 63 pounds if the weights are to be placed in just one pan of the scales.

* 14.[1] Prove that each number can be represented in exactly one way in the following form:

$$a + 3\,b + 9\,c + 27\,d + \cdots$$

where each of a, b, c, d is 1, 0, or -1.

* 15. Determine a set of four weights which make possible weighing loads from 1 to 40 pounds if weights can be placed in either of the pans. How can you generalize this problem?

8. Nim: how to play it.

An interesting application of the number system to the base 2 is the system for winning a game called **Nim.** The theory of this game is complete in that if one knows the system he can always win unless his opponent (either by accident or design) also follows the system, in which case who wins depends solely upon who plays first. The story is told of a well-known intellectual of doubtful morals who, on meeting on shipboard an unsuspecting sucker in the form of one who thought he knew all about the game, used his knowledge so artfully as to completely finance his trip abroad. Let the reader take warning.

In the simplest form of the game a certain number of counters are put into three piles or rows. To play, one draws as many counters as he wishes from a pile; they must all be from the same pile and he must draw at least one. There are two contestants, they play alternately, and the winner is the one who draws the last of the counters. By way of illustration, suppose at a stage in the game there remain only two piles of two counters each. If it is A's turn to play he loses for if he draws a complete pile, B will draw the remaining pile; if he draws one counter from one pile, B can draw one from the other pile and in his following turn

---

[1] See topic 5 at the close of the chapter.

A will have to draw one leaving just one for B. For brevity's sake we can therefore call 0 2 2 a **losing combination** since whoever is faced with that array will lose if his opponent plays correctly. On the other hand, 1 1 1 is a **winning combination.** In general, if it is a contestant's turn to play and the situation is such that he can win by playing properly, no matter how his opponent plays, we say that he is faced with a **winning combination;** if, on the other hand, it is true that, whatever his play, his opponent can win by playing correctly, we call it a **losing combination.** It will turn out that every combination is one or the other of these. The best way to begin the investigation of this game is to play it.

EXERCISES

1. Which of the following are winning combinations: 1 2 3, 1 4 5, 0 4 4, 3 4 5, 2 1 1?

2. Can any of the above be generalized into certain types of winning combinations?

9. Nim: how to win it.

We shall first give the device by which it may be determined, whatever the numbers in the piles, whether any given combination is "winning" or "losing." After that we shall prove that our statement is correct. The method of testing which combination it is is as follows: Write the numbers in the piles in the number system to the base 2 and put them in the position for adding; if the sum of every column is even, it is a losing combination — otherwise it is a winning combination. For instance,

$$3 \quad \text{is} \quad 11$$
$$4 \quad \text{is} \quad 100$$
$$5 \quad \text{is} \quad 101$$

The sums of the first and third columns are 2 but the sum of the second is 1, which is odd. Hence, when the numbers of counters in the piles are 3, 4, and 5 respectively, we have a winning combination. But if 5 were replaced by 7, the last row would be 111 and the test shows that 3, 4, 7 is a losing combination. A direct analysis of the possible plays will show that this is indeed the case. Notice that we *do not add the three numbers to the base 2;* we merely *add each column.*

By way of saving circumlocution, let us call it an **even combination** when, after writing the numbers of the counters in the piles to the base 2, the sum of every column is even; otherwise call it an **odd combination.** We then wish to prove that **an odd combination is winning and an even combination losing.** Notice first that a player can win in one play if there remains just one pile and only under those circumstances can he win in one play. In that case, the number of counters in that pile written to the base 2 will have at least one 1 in it and since the other numbers are 0, it will be an odd combination. Hence a player certainly cannot win in one play unless he is faced with an odd combination. Thus, if a player can so manage that *every* time after he has played, the combination is even, his opponent cannot win; and someone must win since there cannot be more plays than the total number of counters. It, therefore, remains for us to show: (1) that a play always changes an even combination into an odd one, and (2) that faced with an odd combination one can always so play that the result is an even combination. To show the first it is only necessary to see that taking any number of counters from a pile changes the number in that pile written to the base 2; that is, at least one 1 will be changed to a 0 or at least one 0 to a 1 — then the sum of that column if even before the play must be odd after the play. To show the second, pick out the first column, counting from the left, whose sum is odd. At least one of the numbers written to the base 2 must have a 1 in this column. Call one such number, R. It will look like this:

$$\underline{\hspace{3cm}}1\underline{\hspace{3cm}}$$

where the dashes indicate digits (to the base 2) that we do not know about. Find the next column (again counting from the left) whose sum is odd. The digit of R in this column may be 0 or 1. To be noncommittal, call it a. And so proceed until we have R looking like this:

$$\underline{\hspace{2cm}}1\underline{\hspace{1cm}}a\underline{\hspace{2cm}}b\underline{\hspace{1cm}}c\cdots$$

where 1, a, b, c, $\cdots$ are the digits in the columns whose sums are odd. Now write the number T

$$\underline{\hspace{2cm}}0\underline{\hspace{1cm}}A\underline{\hspace{2cm}}B\underline{\hspace{1cm}}C\cdots$$

where A is 0 if a is 1 but 1 if a is 0, similarly for B, C, $\cdots$. In other words, T is the number obtained from R by changing from 0 to 1 or 1 to 0 all the digits of R which occur in columns whose sum is odd. Now T is less than R and hence we can take enough counters from the pile containing R counters to have T left. Then the sum of each column will be even and we have completed our proof.

We illustrate the process for the following example. Suppose we are faced with 58, 51, and 30 counters in the respective piles. We first write these numbers in the binary system

58	is	111010
51	is	110011
30	is	11110
		* ***

where we have placed a star under each column whose sum is odd. R, the pile from which we draw, will be a number which has a 1 in the first starred column counting from the left. In this case all three numbers have this property, so that we may choose R to be any one of the three numbers. Choose it to be the top number, 58. We then change every digit in the top line which occurs in a starred column. The top line becomes 101101 which is 45 in the decimal system and every column will then have an even sum. Hence, to win, take 13 counters from the pile having 58. We could also win by taking 15 from the second or 21 from the third. Notice that if one of the numbers had a 0 in the first starred column counting from the left, we could not draw from that pile to win, for replacing the first 0 by a 1 would increase the size of the number.[1]

EXERCISES

1. Determine which of the following combinations are odd and state in each case of an odd combination how you would play to leave an even combination:

    a. 3 7 9.   *Ans.* Remove 5 from the pile containing 9.

    b. 4 11 29.

    c. 2 4 8.

    d. 4 10 14.

2. Prove that if there are 6, 5, 4 counters in the three piles respectively, one can play in three different ways to leave an even combination.

3. Under what conditions will it be possible to play in three different ways to leave an even combination? Will it ever be possible to play in more than three different ways to leave an even combination? Will it ever be possible to play in exactly two different ways to leave an even combination?

* 4. Sometimes the game of Nim is played so that the *loser* is he who takes the last counter. What would then be the system of winning?

10. Subtraction and division.

You give the cashier one dollar to pay for your thirty-five-cent meal and she drops the coins into your eager hand, trying to make it seem as if you had not spent a thing by counting: "thirty-five, forty, fifty, a dollar." But you, wise as you are, know that she started counting at thirty-five and that another dollar is well on its way toward leaving you forever. It might even be, if you are especially intelligent, that you had already guarded yourself against her sleight-of-hand by answering for yourself the question, "How much change should I get?" As a matter of fact, to ambitious souls, $3 + 7 = 10$ is occasionally interesting, because it is useful in answering the question: "I have 3; how many more do I need to get in order to have 10?" $3 + ? = 10$. It is the process of answering such a question that we call **subtraction.** Assuming we are all ambitious, then, we fortify ourselves by saying that subtraction is a convenient means

[1] See topic 6 at the close of the chapter.

of finding how much more we have to get to achieve our ambition. It goes something like this: I have three dollars and want ten — I am thus three on my way toward having ten and have three less than ten dollars to go. Subtraction is thus reduced to our former method of counting with the difference that we omit some of the objects from our count or, what amounts to the same thing, remove some of the objects before we start to count. Subtracting 3 from 10 is accomplished by taking away three objects from ten and counting what is left. Employing the usual sign for subtraction we have $10 - 3 = 7$ as another way of saying $3 + 7 = 10$.

Subtraction is just the opposite process from addition; for example, if you subtract three and then add three you are just where you started; you merely replace the three you took away. We therefore call **subtraction the inverse of addition.** As we have noted, subtraction is, at least sometimes, a matter of counting but there is this difference from addition — one cannot always prepare the ground for the counting. One cannot take seven things away from three things. To be sure, we can and will define some new numbers so that it can be done in a sense, but the sense is different and we do not consider it now. Then (for positive integers and zero) *subtraction is possible only if the number to be subtracted is less than or the same as the number it is to be subtracted from and under those conditions it is always possible.*

The property of having or not having an inverse has many parallels outside of mathematics. The inverse process is often much more difficult and many times impossible. To put a spoonful of sugar into a cup of coffee is the simplest of acts but to remove it is, if possible at all, a much more difficult process. This is, of course, a kind of addition and subtraction. There is also the idea of opposite or inverse direction. The turnstiles through which one leaves a subway station are expressly built to allow no inverse path. Though the pole vaulter must be careful how he descends, it is not nearly so difficult a feat as propelling himself up and over the bar. It is easier to get into jail than out. The converse of a statement is a kind of inverse if one thinks of it in terms of the direction of implication. For example, consider the two statements: that her title is Mrs. implies that she is a woman; that she is a woman implies that her title is Mrs. Only one of these is true. On the other hand, the following proposition and its converse, or inverse, both hold: that the sum of the squares of the lengths of two sides of a triangle is the square of the length of the third side implies that the triangle is a right triangle. Devotees of *Alice in Wonderland* will remember the following passage:

"Then you should say what you mean," the March Hare went on.

"I do," Alice hastily replied; "at least — at least I mean what I say — that's the same thing, you know."

"Not the same thing a bit!" said the Hatter. "Why, you might just as well say that 'I see what I eat' is the same thing as 'I eat what I see'!"

"You might just as well say," added the March Hare, "that 'I like what I get' is the same thing as 'I get what I like'!"

"You might just as well say," added the Dormouse, who seemed to be talking in his sleep, "that 'I breathe when I sleep' is the same thing as 'I sleep when I breathe'!"

"It *is* the same thing with you," said the Hatter, and here the conversation dropped ....

Apparently the last proposition did have a true inverse for the case considered.

Our definition of "inverse" applies to all of these examples. To put a spoonful of sugar into a cup of coffee and then take it out (without doing anything else) is to maintain the status quo. The pole vaulter starts on the ground and ends on it (though it must be admitted that the exact location is slightly different). If the first proposition and its inverse are combined we get the triviality: that her title is Mrs. implies that her title is Mrs. It should be pointed out that while there is always an inverse proposition there is not always a *true* inverse proposition; similarly one can always describe an inverse action or process but it is not always possible to perform it.

Multiplication also has an inverse process which we call **division.** Sometimes we can divide and sometimes we cannot. Recall that $5 \cdot 6$ means that five sets of six things each are, in all, thirty things; $5 \cdot 6 = ?$ is answered by taking five sets of six dots and counting the total; $5 \cdot ? = 30$ is answered by grouping the thirty objects into five sets with the same number in each set and then counting the number in a set. The existence of such a result, which we call the **quotient,** depends on our ability to group the thirty objects into five sets with the *same* number in each set; that is, depends on our ability to *divide* 30 into five *equal* parts. Notice that the word "equal" is necessary if division is to be the true inverse of multiplication, for in this multiplying we always had a certain number of sets of objects with an *equal* number in each set. Thus 30 can be divided by 1, 2, 3, 5, 6, 10, 15, and 30 but not by any other numbers. To be sure, we later introduce fractions to take care of division in *most* other cases, but, as we have seen in the last paragraph, they involve extensions of our ideas and we shall get along without them for a while. Phrased in terms of counting, "a positive integer b is divisible by c" means that if I have b things I can arrange them in c sets of things with the same number, f, in each set. In other words, when we say that "a positive integer b is divisible by c" or "b can be divided by c" we mean that there is a positive integer f for which $b = cf$.

There are various notations for division. We can write

$$b : c = f, \quad b \div c = f, \quad b/c = f, \quad \text{or} \quad \frac{b}{c} = f.$$

They all mean the same thing and are equivalent to saying $b = cf$. If b is divisible by c we say that c is a **factor** or a **divisor** of b. That which is divided (b in this case) is called the **dividend.** The result of dividing (f in this case) is called the **quotient.** No matter what positive integer the divisor is, there can be no more than one possible value of the quotient.

While all positive integers are divisible by 1, some are divisible by 2, some are divisible by 3, and so forth, but none are divisible by 0 in the sense above defined. This may be seen in at least two different ways. Consider it first in terms of counting: it would be impossible to arrange b things (b being a positive integer) into any number of sets of things with nothing in each. To phrase it in terms of multiplication we may say that if any positive integer b were divisible by 0, then there would need to be a positive integer c such that $b = 0 \cdot c$. But $0 \cdot c = 0$ no matter what positive integer c is and even if $c = 0$. This contradicts our assumption that b, being a positive integer, is not zero.

Of course, this process by which we divide can always be begun if the divisor is not greater than the dividend. From a pile of thirty marbles one could put seven in different places, the eighth with the first, the ninth with the second, and so forth. Finally when we had seven piles of four each with two remaining we would know that 30 is not divisible by 7. We should say that 7 goes into 30 four times with a **remainder** of 2; which is just a brief way of describing the result of our piling. We should have gleaned this information from our labors; that though 30 is not divisible by 7, yet 2 less than 30, which is 28, *is* divisible by 7 — that is the best that can be done. For one number to be divisible by another the remainder must, of course, be zero.

EXERCISES

1. Give various examples outside of mathematics of processes or operations which do or do not have inverses.

2. Is subtraction commutative or associative? Is division?

3. Is division distributive with respect to subtraction?

4. Point out what properties of numbers are used in subtracting 345 from 723.

5. How many possible meanings can the following ambiguous symbol have? 64/8/2? What are the various values?

6. How many possible meanings can the following ambiguous symbol have: 256/16/8/2? What are the various values?

7. In what sense is 0 divisible by every positive integer? If b is a positive integer, to what is $0/b$ equal?

8. Is zero divisible by zero in any sense? Discuss your answer.

9. Examine the process of division of 3240 by 24.

ANSWER: Write

$$24 \mid 3240 \mid 135$$
$$\frac{24}{}$$
$$\frac{84}{}$$
$$\frac{72}{}$$
$$\frac{120}{}$$
$$120$$

In the first line of the division, 2400 is found to be the greatest multiple of 2400 less than 3240. One can omit the 00 after 24 by placing it judiciously below the 32. Again, the 0 in 840 can be omitted for a similar reason, and so it goes; 3240 is found to be $24(100 + 30 + 5) = 135 \cdot 24$.

Explain the process of dividing 3198 by 13.

10. If 3240 and 24 were translated into the dozal notation, would the former be divisible by the latter?

11. If 3240 and 24 are two numbers in the dozal notation (e.g., if 24 is two doza-four), will the former be divisible by the latter?

* 12. a, b, c are three numbers less than 10. When will the two-digit number ab (its digits are a and b in that order) be divisible by c in the dozal notation as well as in the decimal system? Will the two answers ever be the same?

11. Numbers on a circle.

"Tomorrow and tomorrow and tomorrow creeps in its petty pace from day to day." "If winter comes ⋯." Morning, noon, and night, and then another day begins. Sunday, Monday, ⋯, Saturday, and Sunday comes again. "History repeats itself." "Life is a ceaseless round."

Recall the Mad Hatter's tea party with the many places set:

"⋯ It's always six o'clock now."

A bright idea came into Alice's head. "Is that the reason so many tea things are put out here?" she asked.

"Yes, that's it," said the Hatter with a sigh; "it's always tea-time, and we've no time to wash the things between whiles."

"Then you keep moving round, I suppose?" said Alice.

"Exactly so," said the Hatter; "as the things get used up."

"But when you come to the beginning again?" Alice ventured to ask.

"Suppose we change the subject," the March Hare interrupted, yawning, "I'm getting tired of this. I vote the young lady tell us a story."

Alice, of course, had mentioned one of the most important properties of moving around a table; namely, that one must, sooner or later, if one continues in the same direction, come back to the starting point. The table was not round, for it had corners, but as far as the property which Alice mentioned was concerned it might even better have been round. For aesthetic reasons we change the setting slightly and draw a picture of a round table with the seven numbers

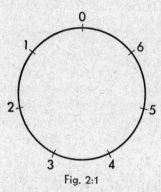

Fig. 2:1

from 0 to 6 instead of teacups. One could even label the divisions of the circle with the days of the week or by increasing the number of divisions and reversing the direction we should have the face of a clock. Motion around the table can be used as a basis for defining a new kind of addition which it is interesting to consider. To add 2, for instance, to any number from 0 to 6, start with the point on the circle bearing the number to which 2 must be added and move two divisions in a counterclockwise direction, that is, in the direction opposite to that in which the hands of a clock move when it is running properly. Thus, to get $3 + 2$ we start at 3 and, moving two divisions in a counterclockwise direction, we come to 5. Hence, even on the circle, $3 + 2 = 5$. Adding by the same method, wonder of wonders, $3 + 4 = 0$ and $4 + 5 = 2$! We call this process **addition on the circle.** No matter what addition we perform, our result is one of the seven numbers. We subtract by merely going in the opposite (that is, clockwise) direction.

We say that the numbers 0, 1, $\cdots$, 6 on the circle and the process of addition on the circle form a **group** because they have the following properties:

1. The result of every addition is one of the numbers — we say the set of numbers is **closed under addition.**
2. Addition is associative.
3. Addition always has an inverse; that is, for any number of the set, a, and any number b (the same or different) there is a number c of the set such that $a + c = b$, or briefly, subtraction is always possible.

The proof that property 2 holds we temporarily postpone. The property 3 is seen by noting that c is the number of divisions of the circle one covers in going in a counterclockwise direction from mark a to mark b. Since addition is commutative in the above example, we say that the set of numbers and the process form an **Abelian group.**

More formally, a group may be defined as follows: Given a set of one or more elements (things to work with, e.g., numbers in this section or angles or permutations in section 15 of this chapter) and a process or operation (such as addition or multiplication) which, to be noncommittal, we denote by $\#$. Then the elements and the process are said to form a group if, for every pair of elements a and b (the same or different) the following properties hold:

1. $a\#b$ is an element of the set: the closure property.
2. $(a\#b)\#c = a\#(b\#c)$: the associative property.
3. There are elements c and d of the set such that $a\#c = b$ and $d\#a = b$: the existence of an inverse.

If, in addition to the above properties, $a\#b = b\#a$, for every pair of elements of the set, it is called an **Abelian group.**

Since we have now given a technical meaning to the term "group" we must be more careful about our use of it in the remainder of this book.

It is also interesting to consider multiplication on a circle. To get $2 \cdot 3$ we can take two steps around the circle (numbered from 0 to 6) taking three divisions in a stride. Since

we should end on the 6 we have the natural result that
$2 \cdot 3 = 6$. But, using the same process, we get the astonishing result that $3 \cdot 4 = 5$. Lo, we need a new multiplication
table! Here it is.

	1	2	3	4	5	6
1	1	2	3	4	5	6
2	2	4	6	1	3	5
3	3	6	2	5	1	4
4	4	1	5	2	6	3
5	5	3	1	6	4	2
6	6	5	4	3	2	1

Notice that though this table could be obtained by counting out on a circle it could also be derived by considering
the remainders when the various products are divided by 7.
For instance, $5 \cdot 3 = 15$ has the remainder 1, which is what
appears in the table as the product of 5 and 3. Notice that,
as in the case of addition, all our results are among the
numbers from 0 to 6.

What of the inverse of multiplication? We use our table.
$5 \cdot ? = 6$. To find the answer, look for the number 6 in the
fifth row of the table and see that $5 \cdot 4 = 6$. (Of course the
fifth column does as well.) That means that if, beginning
on the circle at 0, we take four steps of five divisions each,
we come to rest on 6; or, looking at it in another way, when
$5 \cdot 4 = 20$ is divided by 7, the remainder is 6. That there is
an answer to such a question, that is, that division by any
number from 1 to 6 is always possible, can be seen from the
fact that in each row of the table occur the numbers from
1 to 6 inclusive.

Let us consider two questions about this table:

Question 1: Do the numbers 0, 1, 2, 3, 4, 5, 6 and multiplication on the circle of 7 divisions form a group? The
answer is "no" since the third property of a group fails in
the case of the equation $0 \cdot x = 3$ which has no solution on
the circle.

Question 2: Do the numbers 1, 2, 3, 4, 5, 6 and multiplication on the circle of 7 divisions form a group? The answer

is "yes" since the product of any two numbers of this set is a number of the set; property 2 holds by the proof below; property 3 holds for the following reasons: If a and b are any of the numbers 1 to 6 inclusive, to find the number x for which $ax = b$, look in the ath row of the table for the number b and x will be the number of the column in which b occurs and, since every number from 1 to 6 appears in each row, one can find an x no matter what a and b are. For example, to solve $3x = 4$, look for 4 in the third row, find it in the sixth column and see that $x = 6$ is a solution of the equation on the circle.

The answer to the second question naturally leads to another: In a circle of n divisions, do the numbers 1, 2, $\cdots$, $n - 1$ and multiplication on that circle form a group? (Why, if 0 is included, must they fail to form a group?) If n has two factors, neither of which is 1, we will not have a group since the product of those two factors on the circle will be 0 which is not among the numbers 1, 2, $\cdots$, $n - 1$; for instance if $n = 12$ we have $3 \cdot 4 = 0$. On the other hand, we shall see that *if n has no factors except itself and 1, that is, if it is a* **prime number,** *the numbers 1, 2, $\cdots$, $n - 1$ and multiplication on the circle of n divisions do form a group.* This result could not be proved by examining any amount of tables for various values of n and, even if it were, the computation of a table for $n = 103$, for instance, would be a prodigious task. But without any computation we can prove it for *all* prime numbers n. This is the proof:

 a. The property of closure holds, for the product of any two of the numbers of the set must be one of the numbers 0, 1, 2, $\cdots$, $n - 1$ and it cannot be zero since n is a prime number.

 b. The associative property holds as proved below.

 c. To show that the inverse property holds we must show that each of the numbers of the set occurs in each row. Since each row has $n - 1$ numbers in it, this will be equivalent to showing that no two are equal on the circle. To show that no two are equal we see what would happen if two were equal. That is, we suppose that in the ath row, two numbers, ax and ay, are equal; then their difference would be 0 and, being equal to $a(x - y)$ [see Exercise 16 below], would be in that row. Thus if two numbers in the row were equal, the row would contain a 0, which is not the case. This forces us to conclude that no two numbers in any row are equal and hence that each row contains all numbers 1, 2, $\cdots$, $n - 1$.

We have seen that to find the value of any number on our circle of 7 divisions we divide the number by 7; the number of complete revolutions is the quotient and the value of the number on the circle is the remainder. There are three important consequences of this statement:

1. A number (off the circle) differs from its value on the circle by a multiple of 7.

2. Two numbers whose difference is a multiple of 7 will have the same value on the circle, for in order to get from one to the other on the circle one covers a certain number of complete revolutions.

3. Two numbers which have the same value on the circle must differ by a multiple of 7, for in order to get from one to the other on the circle, one must traverse a certain number of complete revolutions. As a result of these statements we shall demonstrate the following important

Property of multiplication on a circle: To obtain the value on a circle of n divisions of a product of numbers one may first compute the product of their values and find its value on the circle. For example, on the circle of 7 divisions, we wish the value of $8 \cdot 11 \cdot 10$. To do this we should, without using the property, find the product, 880, and divide by 7, leaving the remainder of 5 which is the value on the circle of the product. By using the property we get this more quickly by taking the product of 1, 4, and 3 which are the respective values of 8, 11, and 10 on the circle of 7 divisions. This product is 12 which has the value 5 on the circle. To see why this property holds for a product of two numbers and $n = 7$, let a and b be any two numbers and A and B their respective values on the circle of 7 divisions. Then a differs from A by a multiple of 7 and we can write $a = A + 7r$ and similarly $b = B + 7s$ where r and s are some integers. Thus,

$$ab = (A + 7r)(B + 7s) = A(B + 7s) + 7r(B + 7s)$$
$$= AB + 7sA + 7rB + 49rs$$
$$= AB + 7(sA + rB + 7rs).$$

Hence ab differs from AB by a multiple of 7 and, by statement 2 above, must have the same value on the circle. One could carry through the same process if 7 were replaced by some other number n, and if there were three or more numbers in the product. This would establish our property. It would have as an immediate consequence that *multiplication on a circle is associative*, for $(ab)c = a(bc)$ off the circle and, by the above property, the equation still holds if a, b, and c are replaced by their values on the circle. Similar considerations show that *multiplication on a circle is commutative*.

The above property can be used to find the remainders when certain large numbers are divided by small ones. For instance,

$$8^{20} - 1$$

is a number with 19 digits and to find its value and divide by 7 would be a laborious task; but on the circle with 7 divisions, it has the same value as $1^{20} - 1 = 0$ since 8 has the value 1 on the circle. Hence $8^{20} - 1$ is divisible by 7. To find the remainder when the same number is divided by 5 we use a circle with 5 divisions. Since 8^{20} is 8 taken as a product 20 times, it is equal to 64 taken as a product 10 times, and so proceeding we have

$$8^{20} = 64^{10} = 4^{10} = 16^5 = 1^5 = 1.$$

This then shows that $8^{20} - 1$ is divisible by 5.

It is sometimes useful to consider what might be called *negative numbers on a circle*. Just as we associate the number b with the point reached by taking b steps in a counter-clockwise direction from the point 0, so we can associate $-b$ with the point reached by taking the same number of steps in a clockwise direction. Thus $-b$ and $n - b$ represent the same point on the circle. We define $(-a)(-b)$ to be the same point as $(n - a)(-b)$. It is not hard to show (see Exercise 19 below) that $(-a)(-b) = ab$. This knowledge would have saved a little labor in the last example above since we could then write: $64 = -1$ on the circle with 5 divisions and

$$8^{20} = 64^{10} = (-1)^{10} = 1$$

on the circle with 5 divisions.

Let us consider one further interesting property of multiplication on a circle. First recall that 3^4 means $3 \cdot 3 \cdot 3 \cdot 3$. (The "exponent" 4 tells how many times 3 is taken as a factor.) $3^4 = 81$ has a remainder 4 when divided by 7 and hence on the circle with 7 divisions $3^4 = 4$. A little calculation shows that the remainders of successive powers of 3 are 3, 2, 6, 4, 5, 1. Hence on this circle there is always an answer to the question $3^? = k$ if k is any of the numbers from 1 to 6 inclusive. But the table for the powers of 2 is 2, 4, 1, 2, 4, 1 and the corresponding question does not always have an answer. The interested student will find this subject pursued further in most books on the Theory of Numbers under the heading: "primitive roots." (In general, a number a is called a **primitive root** of b if, on a circle of b divisions, every number except 0 is a power of a; 3 but not 2 is a primitive root of 7.) Every prime number has primitive roots but not all numbers have them.

In the next section we give one application of addition and multiplication on a circle. In section 15 other examples of groups are given.

EXERCISES

1. Would the numbers from 0 to 8 arranged around a circle with the process of addition on the circle form a group?

2. For every positive integer n would it be true that the numbers

from 0 to n arranged around a circle with n divisions form, with the process of addition, a group?

3. Did Magellan's voyage around the world prove that the earth was round?

4. What relationships can you discover between addition of hours on the face of a clock and the dozal system of notation?

5. Form the addition tables for the circles of 7 and 8 divisions. Will these numbers with addition form an Abelian group?

6. Show that the following property of addition holds: to obtain the value on a circle of n divisions of a sum of numbers one may first compute the sum of the values of the numbers on the circle and then find the value on the circle of that sum.

7. Show that addition on a circle is associative.

8. Consider the numbers from 0 to 10 inclusive written on a circle and make the resulting multiplication table. Do the numbers from 1 to 10 and multiplication on the circle form an Abelian group?

9. Write the numbers from 0 to 11 on a circle and consider the questions raised in Exercise 8. How does the multiplication table differ essentially from that for 7 or 11 divisions?

10. For which of the following number of divisions of the circle will multiplication of non-zero numbers yield a group: 13, 16, 15, 26, 31?

11. On a circle of twelve divisions, which of the following are solvable for x? When there is a solution, give all solutions. Where there is no solution, show why there is none.

$$a. \ x + 8 = 3. \qquad b. \ 5x = 7.$$
$$c. \ 2x = 5. \qquad d. \ 10x = 2.$$

12. Find the remainder when 9^{30} is divided by 10, that is, the last digit in the given number. What is the remainder when the same number is divided by 4?

13. Prove that the cube of any positive integer is either divisible by 7, one more than a multiple of 7, or one less than a multiple of 7.

14. Prove that the square of any odd number is one more than a multiple of 8.

15. A dog told a cat that there was a mouse in the five-hundredth barrel. "But," said the cat, "there are only five barrels there." The dog explained that you count like this:

$$1 \quad 2 \quad 3 \quad 4 \quad 5$$
$$9 \quad 8 \quad 7 \quad 6$$
$$10 \quad 11 \quad 12 \quad 13$$

and so on

so that the seventh barrel, for instance, would be the one marked 3, the twelfth barrel the one numbered 4. What was the number on the five-hundredth barrel? (From Dudeney, *Modern Puzzles*.)

16. Prove that on a circle of n divisions multiplication is distributive with respect to addition.

17. In the proof of the associative property on the circle, the number 7 was used. To what extent does this prove the property for a circle of n divisions? Why were the letters used in that proof?

18. Find an example of an indirect proof in the above section and analyze its structure after the manner of Chapter I.

19. Prove that on a circle of n divisions $(-a)(-a) = a^2$.

20. Do the numbers 1, 5, 7, 11 on the circle of 12 divisions and the process of multiplication form a group?

* 21. Notice that in the multiplication table given above both diagonals read the same from left to right as from right to left, the diagonals being 1 4 2 2 4 1 and 6 3 5 5 3 6. Why is this so for other circles? Can you find any other interesting properties of the multiplication tables?

* 22. If b and m have no factors in common except 1, show that if $bx = by$ on the circle of m divisions, then $x = y$ on that circle. Show that this implies that the numbers: b, $2b$, $3b$, $\cdots$, $(m-1)b$ are in some order equal on the circle with m divisions to 1, 2, $\cdots$, $m-1$, if b and m have no factors in common.

* 23. Show that the positive numbers less than m, having no factors in common with m with the process of multiplication on a circle of m divisions, form a group.

* 24. A circle is divided into n parts, where n is a positive integer. Prove that if, for a given positive integer b, m is the smallest number of times one can lay b off on the circumference and have a closed polygon (that is, return to a point previously touched), then n/m is the greatest common divisor of b and n, that is, n/m is the largest number which divides both b and n.

12. Tests for divisibility.[1]

It is easy to tell in any case whether or not subtraction is possible merely by comparing the size of the numbers. But it is more difficult to tell whether division is possible short of actually carrying through the process to see how it comes out. However, in some cases, one can easily tell ahead of time whether or not division is possible. This is easiest for 2, 5, and 10. There also is a well-known test for divisibility by 9. To illustrate this, consider the number 145,764. Add the digits, that is, 1, 4, 5, 7, 6, and 4 to get 27. This test tells us that since this sum is divisible by 9, the

[1] See topic 8 at the end of the chapter.

number itself is divisible by 9. If the sum had not been divisible by 9 the number given would not have been divisible by 9. The test then is this: *it is true that a number is divisible by 9 if the sum of its digits is divisible by 9 and only in that case.*

It is not hard to prove this rule. Let us see how it goes for the number 1457. This number can be written

$$1000 + 400 + 50 + 7 = 1000 + 4(100) + 5(10) + 7.$$

Using the result of Exercise 6 of the last section for $n = 9$, together with the corresponding property for multiplication, we see that the value of the sum on the circle of 9 divisions, that is, its remainder when divided by 9, is the same as the value of $1 + 4 + 5 + 7$ since 1000, 100, and 10 all have the remainder 1 when divided by 9, that is, the value of 1 on the circle of 9 divisions. Notice that we have proved for this number more than we set out to prove, that is, *the remainder when the sum of the digits is divided by 9 is the same as the remainder when the given number is divided by 9.*

The proof for any number of four digits would go as follows. Suppose the digits are in order a, b, c, d. Then the number is equal to

$$1000\,a + 100\,b + 10\,c + d$$

which has the same value on the circle of 9 divisions as $a + b + c + d$. We have thus proved the latter italicized statement above for four-digit numbers. Another way of saying the same thing is that the value of a number on the circle with 9 divisions is the value of the sum of its digits.

This result, which holds for every number, has a useful application in the process which goes by the name of "casting out the nines" [1] — an excellent and ancient way of checking addition and, for that matter, subtraction, multiplication, and division. For purposes of illustration consider the sum

[1] See reference **38.**

$$137$$
$$342$$
$$\underline{890}$$
$$1369$$

We know that the value of the sum on the circle of 9 divisions is the same as the value of the sum of the values. The sum $1 + 3 + 7 = 11$ and (being very lazy) $1 + 1 = 2$; hence 137 has the value 2 on the circle of 9 divisions. Similarly 342 and 890 have the values 0 and 8 respectively. The sum $2 + 0 + 8 = 10$ has the value 1 which should be the same as the value of 1369, that is, of $1 + 3 + 6 + 9$, which has the value 1 on the circle. The check could be written like this:

137	2
342	0
890	8
1369	1

In practice the process can be shortened by dropping out multiples of 9 as one goes along (this was the reason for the name of the process). For instance, in the above 890 one can omit the 9 and in the sum we drop out 3, 6, and 9 since $3 + 6 = 9$.

EXERCISES

1. Prove that for every number of five digits it is true that the remainder when it is divided by 9 is the same as the remainder when the sum of its digits is divided by 9.

2. Show that by adding the digits of the number, then adding the digits of the sum of the digits, and so on, it is true that to find the remainder when any number is divided by 9 the only division necessary is dividing 9 by 9.

3. Give a test for divisibility by 4.

4. Is there any way to tell solely by the last digit in a number whether or not the number is divisible by 9? By 7? Give your reasons.

5. What is a test for divisibility by 11? Is there any way of comparing the remainders when the number itself and the sum of its digits is divided by 11, as was done previously for 9? If not, can the sum of the digits be replaced by some other simple expression in terms of the digits

to give the desired result? Prove your statement for any number of four digits.

6. In a number system to the base twelve, for what number b would the following be true: a number is divisible by b if the sum of its digits is divisible by b and only in that case.

7. In the previous exercise would there be more than one number b for which the statement could be made?

8. In a number system to the base thirteen, for what numbers b could the statement in Exercise 6 be made?

9. Show that the remainder when $a + b$ is divided by 37 is equal to the remainder when $1000\,a + b$ is divided by 37. For example: $457 + 324$ and 457,324 have the same remainders when divided by 37.

10. Show that the remainder when $b - a$ is divided by 7 is equal to the remainder when $b + 1000\,a$ is divided by 7. For example, $457 - 324$ and 324,457 have the same remainders when divided by 7.

11. Can you find any other numbers which have simple tests for divisibility?

12. What is a test for divisibility by 2 in the number system to the base 3?

13. Check the result of multiplying 257 by 26 by casting out the nines. Show why it works.

14. If a set of numbers is correctly added will it always check under casting out the nines? If a sum checks by casting out the nines is the addition necessarily correct?

15. For the dozal system of notation what would be the check corresponding to casting out the nines?

16. The result for 1457 in the second paragraph of this section could have been obtained more quickly by dividing 1457 by 9 and finding the remainder. Why was this not done?

17. Show how the following trick can be performed and why it works. Also mention any possibility of failure. You select any number, form from it another number containing the same digits in a different order, subtract the smaller of the two numbers from the larger, and then tell me all but one of the digits of the result. I can then (perhaps) tell you what is the other digit of the result.

* 18. If N is any number of four digits (that is, between 1000 and 10,000) show that the sum of the digits of $99\,N$ is 18, 27, or 36. This may be made the basis of a trick as follows. Memorize the 18th name on page 18, the 27th name on page 27, etc., of your local telephone directory. Then announce to a friend that you have memorized the telephone directory and by way of proving it ask him to select some number unknown to you, multiply it by 100, subtract the number from the result, and add the digits of the final result. He tells you his answer and you tell him

he can find such and such a name in, say, the 27th place on page 27. One variation of this trick is not to ask for the number but to find for yourself which it is by asking if it is odd or even, less than 30 or more than 30.

* 19. Can you devise any other tricks similar to that in Exercise 17?

13. Divisors.

We have previously called attention to the fact that the base twelve is convenient for certain purposes because it has, for its size, a large number of divisors. It has, including 1 and 12, six divisors (that is, 12 is divisible by six different numbers). All the numbers through 23 have less than seven divisors. Again, 24 has more divisors than any smaller number. (Such numbers are called **highly composite.**) What is the next such number? We leave the question to anyone who is interested.

Another line of inquiry which has to do with divisors concerns the so-called perfect numbers: $6 = 1 + 2 + 3$ is the sum of its divisors including 1 but not including 6; 6 is thus called a **perfect number.** The next one is $28 = 1 + 2 + 4 + 7 + 14$. Is there another one less than 50? There is a formula which gives all the even perfect numbers [1] but no one has found an odd one. In fact whether or not there are odd perfect numbers is one of the unsolved mysteries of the mathematical world. Fame awaits him who can find an odd perfect number or prove that there are none. The best result so far about odd perfect numbers is that any such number must have at least five different prime [2] factors.

14. Prime numbers.

From the point of view of having divisors, there are numbers that are especially peculiar: integers greater than 1 which have no divisors except themselves and 1. These are the **prime numbers** — the first five are 2, 3, 5, 7, and 11. Integers greater than 1 which are not prime numbers are called **composite numbers.** The worst of it is (though from

[1] See topic 10 at the close of the chapter.
[2] See Section 14.

many points of view it is very fortunate) that *there is no last prime number*. Euclid, of geometry fame, proved this a long time ago. Here is his proof essentially. Suppose the fiftieth prime number were the last one. Multiply together the fifty prime numbers and add 1 to the product. Call the result B. B cannot be divisible by any one of the fifty primes because the remainder is 1 when it is divided by any of them. Call q the smallest number greater than 1 which divides B. If q had a divisor greater than 1 and smaller than q, this divisor would divide B and contradict our assumption about q. Therefore, since q divides B and has no divisors greater than 1 and less than q, q is a prime number different from the fifty primes. Hence our assumption that there are only fifty primes is false. Notice that we have not proved that B or even q is the fifty-first prime but we have proved that q is a prime which is not one of the first fifty. Furthermore, no matter how many primes we thought there were, we could always prove, by the above method, that there is one more. A short way of saying this is: *there are infinitely many prime numbers*. This merely means that no matter what number you name, there are more than that many primes. It is like the various word battles of our childhood: to "My father gets ten dollars every day" you countered with "My father gets a hundred dollars every day." Your opponent in turn raised the ante and whoever could count the furthest won. An interesting thing about the phenomenon in this connection is that if you were boasting about the number of primes and I were boasting about the number of positive integers, again the victory would go to him who could count furthest even though there are many positive integers which are not prime numbers and every prime number is a positive integer. Furthermore, if we extend our concept of "just as many" which we considered in the first section of this chapter, we can say that there are just as many prime numbers as there are positive integers because we can establish a one-to-one correspondence between the prime numbers and the positive integers: there is a tenth prime, there is a millionth prime, there is a trillionth prime, and so on.

There is an interesting theorem about primes which goes by the name of **Bertrand's postulate**. It is this: strictly between any number A ($A > 1$) and its double there is a prime number; where "strictly between" means "between and not equal to A or $2A$," that is, not only greater than A but less than $2A$. For instance, if $A = 5$, then 7 (but not 5) is strictly between 5 and 10. Though it seems not strange in the least that this theorem should be true, it is rather difficult to prove.

At this point we shall mention two statements about the occurrence of prime numbers which most persons believe to be true but which no one has yet been able to prove. Two prime numbers which differ by 2 (such as 29 and 31) are called **twin primes**. It is *believed* that there are infinitely many twin primes. The second unproved statement is this: every even number greater than 2 is the sum of two prime numbers. This latter is called **Goldbach's theorem** but no one has proved it.

Though primes are peculiar from the point of view of divisibility, this very fact makes them the indivisible numbers into which numbers can be factored, for when, in the process of dividing, one comes to a prime, the division stops. *If a number is expressed in two different ways as a product of prime numbers, the two products differ at most in the order in which the factors occur;* that is, each prime factor appears the same number of times in each product. This is called the **law of unique decomposition** into prime factors. For example, consider the number 315. It has a factor 5 and $315 = 5 \cdot 63$. The number 63 is divisible by 3 and $315 = 5 \cdot 3 \cdot 21$. The number 21 is divisible by 3 and $315 = 5 \cdot 3 \cdot 3 \cdot 7$, and all the factors are prime numbers. If we had divided first by 3 the result might have been $315 = 3 \cdot 7 \cdot 5 \cdot 3$. In both of the expressions for 315 as a product of prime factors each prime 3, 5, and 7 appears the

same number of times. A convenient way of expressing the
factorization is $315 = 3^2 \cdot 5 \cdot 7$.

EXERCISES

1. If B is 1 more than the product of the first n primes, find the smallest prime divisor of B. Is it the $(n + 1)$th prime for $a = 3$, for $a = 4$,
for $a = 5$, for $a = 6$?

2. How many prime numbers are there strictly between 11 and 22?

3. We have shown that there are just as many prime numbers as
positive integers. Why does this not prove that every positive integer
is a prime number?

4. Why does Bertrand's postulate imply that there are infinitely many
prime numbers?

5. Prove that there are more than twice as many positive integers less
than 1000 as there are prime numbers less than 1000. Would this be true
if 1000 were replaced by 1,000,000?

6. Prove by use of Bertrand's postulate or by any other means the
following: 1 more than the product of the first n primes is *never* the
$(n + 1)$st prime unless $n = 1$.

7. In 1937, a man by the name of Vinogradoff proved the following
result: there is a number M such that all odd numbers greater than M
are the sum of three prime numbers (e.g., $15 = 3 + 5 + 7$). Would this
result imply the following theorem or would it be implied by it: there is
a number M such that every even number greater than M is the sum of
two prime numbers?

* 8. Show that if A and B have no factor except 1 in common, the
number of divisors of AB is the product of the number of divisors of
A and B. (The number itself and 1 are counted as divisors.)

15. Groups.

Groups consisting of multiplication or addition on a circle
are not by any means the only kinds of groups. In this section we shall discuss other groups one of which is not Abelian.

First recall that the addition table of numbers on a circle
of three divisions looks like this.

	0	1	2
0	0	1	2
1	1	2	0
2	2	0	1

Call this group $A_1(3)$.

Next consider the rotations about its center of the equilateral triangle in the figure. We confine our attention to the three
rotations which take the triangle into coincidence with itself,
that is, more briefly, "take the triangle into itself." These
rotations we call A, B, and C where A stands for no rotation, B a rotation of 120°, and C a rota-
tion of 240° all in a counterclockwise
direction. We define the "sum" of two
rotations to be the result of performing
first one and then the other. For instance,
$B + C$ means "rotate through 120° and
then through 240°." This has the same
result as not moving the triangle at all.

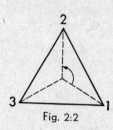

Fig. 2:2

Hence we say that $B + C = A$. The addition table is

	A	B	C
A	A	B	C
B	B	C	A
C	C	A	B

Let us test this for the properties of a group. Since the sum
of two rotations which take the triangle into itself takes the
triangle into itself (or since only A B C occur in the body of
the addition table) the property of closure holds. It follows
from geometrical considerations (or from the fact that each
letter appears in every row and column) that there is always
an inverse to addition, that is, that subtraction is always
possible. The associative property holds. Hence we have a
group. Furthermore, it may be seen that it is an Abelian
group. Call it $A_2(3)$.

Though these groups had different origins it is easy to
see that if we replace A by 0, B by 1, and C by 2 in the
second addition table we get the first table. Any two groups
which have the same addition tables except for the letters
or symbols involved are called **isomorphic.** Two isomorphic
groups are essentially the same. In this case it may be seen
merely by connecting by lines the successive divisions of the
circle for $A_1(3)$ and replacing the "3" by "0." Adding num-

bers on that circle is essentially the same as rotating through
$0°$, $120°$, and $240°$. As a matter of fact, noticing such a relationship in the beginning (perhaps you did) would have
eliminated the necessity of showing that in both cases the
properties of a group hold.

Now we develop a third group, $A_3(3)$, which, in spite of
its different origin, turns out to be isomorphic with these
two. Write

$$a = \begin{pmatrix} 1 & 2 & 3 \\ 1 & 2 & 3 \end{pmatrix} \qquad b = \begin{pmatrix} 1 & 2 & 3 \\ 3 & 1 & 2 \end{pmatrix} \qquad c = \begin{pmatrix} 1 & 2 & 3 \\ 2 & 3 & 1 \end{pmatrix}$$

where a means that we replace 1 by 1, 2 by 2, and 3 by 3;
b means that we replace 1, 2, 3 by 3, 1, 2 respectively and
c means replace 1, 2, 3 by 2, 3, 1 respectively. We cannot call a, b, and c "numbers" and hence we speak of them
as **elements** of the group. Each element is then a reörder-
ing or **permutation** of the numbers 1, 2, 3. The "product"
of any two elements we understand to be that permutation
which results from performing first one and then the other.
For instance, b takes 1, 2, 3 into 3, 1, 2 respectively, and
c takes 3, 1, 2 into 1, 2, 3 respectively. Thus the net result
of applying first b and then c is to leave each number unaltered and we say that $bc = a$. In this fashion we can
form the multiplication table

	a	b	c
a	a	b	c
b	b	c	a
c	c	a	b

where the first line gives the products a^2, ab, ac in order,
etc. We could verify in detail that we have a group but
looking at the previous addition table we see that it is the
same except that capital letters are used instead of small
letters and addition occurs instead of multiplication. Hence
this group is isomorphic to each of the preceding groups.
The geometrical relationship between this group and the
preceding one may be seen by numbering the vertices of
the triangle as in figure 2:2 and seeing, for example, that

B replaces 1 by 3, 2 by 1, and 3 by 2; b does the same thing. The element a behaves like 1 in the multiplication table and corresponds to A which behaves like 0 in the addition table.

In fact, every group with just three elements is isomorphic to every other group with just three elements. The proof of this fact is made easy if we first prove the following property of all groups:

Every group contains an **identity element,** that is, considering the process to be multiplication, an element i such that $ix = xi = x$ for every element x of the group. (What is the identity element when the process is addition?)

PROOF: Let a be some element of the group. From the existence of an inverse, we know that there is some element i such that $ai = a$. Then let b be any other element of the group. There is an element y such that $b = ya$ and from the associative property, $bi = yai = ya = b$. Hence $bi = b$ for every element b. In a similar fashion we can prove the existence of a j such that $jb = b$ for every element b. Replacing b by i in the last equation we have $ji = i$. But we know that $ai = a$ for every element of the group. This shows that $j = i$ and our proof is complete.

To show that every group with three elements is isomorphic to one of these above we designate the three elements by i, r, and s where i is the identity element. We know the following part of the multiplication table:

	i	r	s
i	i	r	s
r	r		
s	s		

Keeping in mind the restriction that each row and column must contain i, r, s in some order we see that there is only one way in which to fill in the blank part of the table, for s cannot occur in the last place in the second row and hence must occur in the second with i in the last. This determines the second and hence last row and we have the multiplication table

	i	r	s
i	i	r	s
r	r	s	i
s	s	i	r

which, except for the letters involved, is the same as the preceding ones.

Notice now that rotations are not the only ways of taking our equilateral triangle into itself. We might keep one vertex fixed and interchange the other two: a rotation about an altitude. This will give us three other elements which can also be represented as permutations in the following manner:

$$d = \begin{pmatrix} 1 & 2 & 3 \\ 1 & 3 & 2 \end{pmatrix} \qquad e = \begin{pmatrix} 1 & 2 & 3 \\ 3 & 2 & 1 \end{pmatrix} \qquad f = \begin{pmatrix} 1 & 2 & 3 \\ 2 & 1 & 3 \end{pmatrix}$$

The following multiplication table you should check at least in part.

	a	b	c	d	e	f
a	a	b	c	d	e	f
b	b	c	a	f	d	e
c	c	a	b	e	f	d
d	d	e	f	a	b	c
e	e	f	d	c	a	b
f	f	d	e	b	c	a

We may test the various group properties geometrically or by the multiplication table assuming, as usual, that the associative property holds. Thus, we may call this a group, G(3). Notice that this group is *not Abelian* since, for example, $dc = f$ but $cd = e$.

If we select the elements a, b, c from G(3) we have the group $A_3(3)$ which is called a **subgroup** of G(3). The elements a and f form another subgroup of G(3). However, the elements a, b, c, d do not form a group because cd is not one of a, b, c, d and the closure property thus fails to hold. It is a property of groups which we shall not attempt to prove here,[1] that the number of elements in any subgroup

[1] See topic 11 at the close of the chapter.

is a divisor of the number of elements in the group. Notice that 2 and 3 are divisors of 6, but 4 is not. Also, since every group contains an identity element, every subgroup must contain the identity element of the group.

Notice that G(3) cannot be isomorphic to the group obtained by adding numbers on a circle with 6 divisions since G(3) is not Abelian and the group on the circle is Abelian.

EXERCISES

1. Prove that the elements of G(3) can be written:

$$a, b, b^2, d, db, db^2$$

and that the multiplication table can be constructed from the following: $b^3 = a$, $d^2 = a$, $bd = db^2$, and the knowledge that the associative law holds. (The elements b and d are called **generators** of the group.)

2. Find the multiplication table of each of the following permutation groups:

1) $a = \begin{pmatrix} 1 & 2 & 3 & 4 \\ 1 & 2 & 3 & 4 \end{pmatrix}$ $b = \begin{pmatrix} 1 & 2 & 3 & 4 \\ 2 & 1 & 4 & 3 \end{pmatrix}$ $c = \begin{pmatrix} 1 & 2 & 3 & 4 \\ 3 & 4 & 1 & 2 \end{pmatrix}$ $d = \begin{pmatrix} 1 & 2 & 3 & 4 \\ 4 & 3 & 2 & 1 \end{pmatrix}$

2) $a = \begin{pmatrix} 1 & 2 & 3 & 4 \\ 1 & 2 & 3 & 4 \end{pmatrix}$ $b = \begin{pmatrix} 1 & 2 & 3 & 4 \\ 2 & 3 & 4 & 1 \end{pmatrix}$ $c = \begin{pmatrix} 1 & 2 & 3 & 4 \\ 3 & 4 & 1 & 2 \end{pmatrix}$ $d = \begin{pmatrix} 1 & 2 & 3 & 4 \\ 4 & 1 & 2 & 3 \end{pmatrix}$

Show that each of the groups above has the closure property and that division is always possible. Point out which are Abelian.

3. There is a connection between each of the groups of Exercise 2 and rotations of a square, analogous to the rotations of a triangle. Show what the connection is.

4. H is the group formed by the numbers of a circle with 6 divisions and the process of addition. K is the group formed by the numbers 1, 2, 4, 5, 7, 8 on a circle with 9 divisions and the process of multiplication. Are H and K isomorphic? Is K isomorphic to G(3)?

5. Is either of the groups of Exercise 2 isomorphic with multiplication or addition on a properly chosen circle?

6. Find all the subgroups of G(3).

7. Show that all groups containing exactly four elements are isomorphic to one of the groups of Exercise 2.

8. A group has the elements: $i, v, u, u^2, u^3, vu, vu^2, vu^3$ where i is the identity element, $u^4 = i = v^2$ and $vu = u^3v$. Write the complete multiplication table and point out any relationships you can find between it and the groups of Exercise 2. Is there any relationship like those indicated in exercises 3 or 5?

16. Topics for further study.[1]

(Where no reference is given, you should not conclude that there are no references but merely that you should have more profit and pleasure from investigating the topic without consulting references.)

1. The history of the number concept and the number system: references **9**, Chaps. 1, 2; **32**, Chaps. 1, 2.

2. Counting and big numbers: references **25**, Chap. 2; **8**, Chap. 17.

3. Numbers to other bases: base 12, references **2, 3, 33**; base 8, references **24, 34**; base 2, reference **25**, pp. 165–171 (puzzles).

4. The Window reader (another way to perform a trick like that at the close of section 7): reference **5**, pp. 321–323.

5. Bachet weights problem (see Exercise 13 of section 7 of this chapter): reference **5**, pp. 50–53.

6. Nim: reference **35**, pp. 15–19.

7. Some properties of powers of numbers on a circle (see section 11).

8. Tests for divisibility: references **28**, Chap. 1; **7**.

9. Highly composite numbers — what ones are they? (See section 14.)

10. Perfect numbers: reference **35**, pp. 80–82.

11. Groups: references **31**, Chap. 17; **6**, pp. 278–283.

With the equipment given in this chapter, any of the following may be accomplished:

a. Prove: If a is an element of a group and t is the least power of a such that $a^t = i$, where i is the identity element, then t is a divisor of the number of elements of the group.

b. Find all the groups (no two isomorphic) with five elements. What ones of these can be obtained by multiplication or addition on a circle?

c. Prove: If a is a fixed element of a group, then the set of elements y of the group for which $ay = ya$, themselves form a group.

---

[1] Boldface references are to the Bibliography at the end of the book.

d. Find three groups with eight elements (there are five), no two isomorphic.

12. Magic squares: references **35,** pp. 159–172; **28,** Chap. VII; **5,** Chap. 7.

13. Puzzles: **5,** Chap. 1; **26; 25,** Chap. 5 (see also further references at the close of that chapter).

14. Find every number whose square is a four-digit number having the last three digits equal, that is, of the form *baaa.*

15. Find a method for solving such problems as the following: What is the smallest number which is equal to 2 on the circle of three divisions, 5 on the circle of 7 divisions, and 8 on the circle of 11 divisions?

16. The Eight Queens Problem (the problem is to place eight queens on a chess board so that no one will threaten any other one): references **26,** Chap. X; **18.**

17. A "miniature geometry" based on the set of numbers on a circle: reference **31,** pp. 448–453.

18. What mathematicians think about themselves: references **6, 21, 37.**

CHAPTER III

Negative Integers, Rational and Irrational Numbers

1. Inventions.

The numbers 0, 1, 2, 3, ⋯ used throughout Chapter II are called **nonnegative integers;** 0 is zero, and 1, 2, 3, 4, ⋯ are the positive integers. These numbers, the decimal system for the representation of them, and the rules for computing with them, were created by human ingenuity to meet various human needs. To assume that men have always known and been able to use the results of Chapter II is like assuming that they have always had motorcycles and been able to ride them. The fact is that numbers and vehicles alike are not mysterious creations of nature plucked from trees or hooked in lakes; they are "of the people, by the people, and for the people." Just as there are several kinds of vehicles (oxcart, wagon, buggy, train, steamship, motorcycle, automobile, submarine, dirigible, airplane), so there are several kinds of numbers (zero, positive integers, negative integers, primes, perfect numbers, rational numbers, algebraic numbers, irrational numbers, real numbers, imaginary numbers, complex numbers). Just as new kinds of vehicles have been invented to have new uses, new kinds of numbers have been invented to have new uses. In this chapter we first introduce negative integers and find that the set of all integers (which includes 0, the positive integers, and the negative integers) has an important property which the smaller set of positive integers fails to have; namely, that, for every pair of integers a and b, there is an integer x such that $a + x = b$.

Alice has something to say on the subject.

"And how many hours a day did you do lessons?" said Alice, in a hurry to change the subject.

"Ten hours the first day," said the Mock Turtle; "nine the next, and so on."

"What a curious plan!" exclaimed Alice.

"That's the reason they're called lessons," the Gryphon remarked, "because they lessen from day to day."

This was quite a new idea to Alice, and she thought it over a little before she made her next remark. "Then the eleventh day must have been a holiday?"

"Of course it was," said the Mock Turtle.

"And how did you manage on the twelfth?" Alice went on eagerly.

"That's enough about lessons," the Gryphon interrupted in a very decided tone; "tell her something about the games now."

Our immediate association of negative numbers with the above quotation bears witness to the fact that we are so accustomed to various *interpretations* of negative numbers that we are apt to forget that their origin is really an *invention*. But we are getting ahead of our story.

2. Negative integers.

It was pointed out in Chapter II that if each of a and b is one of the numbers $0, 1, 2, \cdots$ and a is not greater than b, then one of the numbers $0, 1, 2, \cdots$ (say x) is such that

$$a + x = b.$$

For example, if $a = 3$, $b = 7$, then $x = 4$. If, however, a is greater than b, then it is impossible to choose x from the numbers $0, 1, 2, \cdots$ in such a way as to make

$$a + x = b.$$

For example, if $a = 1$, $b = 0$, then

$$1 + x = 0$$

is an equation which cannot be true when x has one of the values $0, 1, 2, \cdots$. Likewise there is no value for x among the numbers $0, 1, 2, \cdots$ which can satisfy any of the equations

$$1 + x = 0; \quad 2 + x = 0; \quad 3 + x = 0 \cdots.$$

You may ask: "Is there a number x such that $1 + x = 0$?" If by "number" you mean one of the numbers 0, 1, 2, $\cdots$, then the answer is "no." But we may take advantage of the inventive genius of the daring human who said "Let there be a number x such that $1 + x = 0$, let the name of this x be *minus one*, and let this x be denoted by the symbol -1." We accept this creation of -1 and likewise create -2, -3, -4, $\cdots$ such that

$$1 + (-1) = 0; \quad 2 + (-2) = 0; \quad 3 + (-3) = 0 \cdots.$$

We now have not only 0 and the positive integers 1, 2, 3, $\cdots$, but also the **negative integers** -1, -2, -3, $\cdots$.

Of course it is true that our various interpretations of negative numbers arose out of just such a need as gave rise to their creation. The notion of a negative number is a kind of unifying idea back of many situations which are described in many different terms. We say, "it is five below zero" when we mean that the temperature must rise by five degrees before it is at zero; in accordance with our definition we interpret $-5°$ to have the same meaning. In some games we can be "ten in the hole" (and often write it ⑩) when we need ten points to have the score we began with; in accordance with our definition we interpret -10 to have the same meaning. We say "we are four dollars in the red" when we need to add four dollars to our capital to be square with the world; in accordance with our definition we interpret -4 to have the same meaning. In all these situations it is a matter of interpretation. Without such an interpretation, to say "I have minus four dollars" has no more meaning than to say "I have four red dollars."

3. The integers.

The whole set of integers, which includes 0, the positive integers, and the negative integers, is often displayed in the array $\cdots$, -3, -2, -1, 0, 1, 2, 3, $\cdots$. This array is familiar to everyone who has seen a thermometer turned sidewise. A schematic picture is shown on the following page.

It is convenient to call the point above the number 7, **the point with coordinate 7** or, briefly, **the point 7**; in general, the point in the figure above an integer n is the point with coordinate n, or the point n. The process of adding 4 to 3 in order to get 7 may be associated with the process of starting 3 units to the right of 0 and then going 4 units to the right, thus reaching a point 7 units to the right of 0.

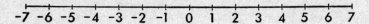

The act of subtracting 2 from 5 to get 3 may be associated with the process of starting 5 units to the right of 0 and then going to the left 2 units, thus reaching a point 3 units to the right of 0. To phrase this statement in general terms: *Addition is associated with motion to the right on a line and its inverse, subtraction, with motion to the left on a line, just as motion in a counterclockwise direction on a circle is associated with addition and clockwise motion with subtraction.*

In the first chapter we found that $3 - 3 = 0$. Here we have defined -3 so that $3 + (-3) = 0$. Furthermore, not only is $5 - 3 = 2$ but $5 + (-3) = 2 + 3 + (-3) = 2$. Similarly, assuming that addition is associative for all integers, it may be seen that subtracting 3 has the same effect as adding (-3). In fact, if b is any positive integer, subtracting b can be shown to have the same effect as adding $(-b)$. Hence we find it convenient to say that adding a negative number $(-b)$, that is, moving to the right $(-b)$ units, shall mean moving b units to the *left* on the line. It is also convenient to say that subtracting $(-b)$ shall be equivalent to adding b if b is some positive number, that is, moving to the left a negative number of units shall mean (and this is a *definition*) moving to the right the corresponding positive number of units. For example:

$$5 - (-3) = 5 + 3 = 8.$$

From the above paragraph it follows that whatever integers a and b are, $a + b$ will be an integer for we start at the point a and "go to the right" b units, the actual motion being to the left if b is negative. Similarly $a - b$ is obtained by "going to the left" b units, the actual motion being to the right if b is negative.

Although we do not attempt to give a proof, it is fairly evident from the idea of "addition on a line" that the equations

$$a + b = b + a$$
$$(a + b) + c = a + (b + c)$$

hold whenever a, b, and c are integers.

EXERCISES

1. What, in terms of the thermometer placed on its side, are the interpretations of the associative and commutative laws of addition?

2. If the temperature at 6:00 P.M. is $+5°$ and the temperature falls $8°$ between 6:00 and midnight, what is the temperature at midnight?

3. What is the distinction graphically (that is, on the thermometer) between $-2 + 5$ and $5 - 2$? Express in letters a property of integers which shows that $-a + b = b + (-a)$.

4. Show that $(2 + 7) - 8 = 2 + (7 - 8)$ and that $2 - (8 - 7) = (2 - 8) + 7$. Express in letters the properties of which these are examples.

5. Is part or all of the following statement true: If a and b are positive integers, $a - b$ as well as $a + b$ are positive integers?

6. Show that the set of all integers, with the process of addition, form an Abelian group. How does this group differ from the types of groups discussed in the first chapter?

7. Does the set of all integers with the process of multiplication form a group?

4. Multiplication of integers.

We found in Chapter II that if a and b are positive integers, then the product ab can be visualized as the total number of objects in a sets of which each set contains b things. The question of evaluating the product ab when one (or both) of a and b is negative is more delicate. As an actual matter of fact, such products have not yet been defined in this course and they are entirely meaningless until meaning has, in some way or other, been attached to them. The simplest thing for us to do here is to agree that if c and d are both nonnegative integers, then the product $c \cdot d$ shall be as defined in Chapter II and that

$$(-c) \cdot d = -(c \cdot d)$$
$$c \cdot (-d) = -(c \cdot d)$$
$$(-c) \cdot (-d) = c \cdot d.$$

Thus, for example, $(-2) \cdot 3 = -6$, $3 \cdot (-4) = -12$ and $(-5) \cdot (-5) = 25$. For brevity's sake we write $-(3 \cdot 5)$ as $-3 \cdot 5$. As in the case of addition $2 - 3 \cdot 5$ means: multiply 3 by 5 and subtract the result from 2.

The distributive property holds also for negative integers. Suppose, for instance, a, b, and d are nonnegative integers. We can show

$$a(b - c) = ab - ac$$

in the following fashion. Let $b-c=d$. This means $b=c+d$. Multiply this by a to get $ab = a(c + d) = ac + ad$ from the distributive property for nonnegative integers. But $ab = ac + ad$ means $ab - ac = ad$ and the latter is equal to $a(b - c)$. Similarly, to show

$$-a(b + c) = -ab - ac$$

we notice that $-a(b + c)$ is defined as that number which, when added to $a(b + c)$, yields zero. But $a(b + c) = ab + ac$ and if $-ab - ac$ is added to it, the result is zero. This shows what we wished to show. These should make plausible the fact that

$$a(b + c) = ab + ac$$

for any integers a, b, and c negative as well as positive or zero.

EXERCISES

1. Carefully use the definitions of products to show that
$$-3(2 - 7) = (-3)2 + (-3)(-7).$$

2. In each of the following expressions insert one pair of parentheses in as many ways as possible (there are 15 ways) and evaluate the resulting expressions:

   $a.$ $1 - 2 - 3 - 4 - 5,$
   $b.$ $1 - 2 \cdot 3 - 4 - 5.$

How many different results are there? Which are equal to the expression written without any parentheses?

3. Pick out all correct right-hand sides for each of the following equations and give reasons for your answer: ·

$$a. \quad Sr - S = \begin{cases} r(S - 1) \\ S(r - 1) \\ r(S - 1) + r - S \end{cases}$$

$$b. \quad ax - ay + bx - by = \begin{cases} (a - b)(x + y) \\ (a + b)(x - y) \\ (a + x)(b - y) \end{cases}$$

$$c. \quad ab - ac + bc = \begin{cases} a(b - c) + bc \\ c(a - b) + ab \\ b(a - c) + ac \\ -b(-a - c) - ac \end{cases}$$

4. Considering all possible cases in which neither one nor both of a and b are negative, prove that

$$ab = ba$$

when a and b are integers. You may use the result of Chapter II that $cd = dc$ when c and d are nonnegative integers.

5. Prove that $(-ab)c = -a(bc)$ where a, b, and c are nonnegative integers assuming the associative property of multiplication for nonnegative integers.

6. Assume that a, b, c are nonnegative integers. Using the results of Chapter II and the definitions of this chapter, prove

$$a(-b - c) = -ab - ac.$$

5. Introduction of rational numbers.[1]

If a, b, and x are integers such that $a \neq 0$ (i.e., a is different from 0) and

$$ax = b$$

then it is customary to write

$$x = \frac{b}{a}, \quad x = b/a, \quad x = b : a, \quad \text{or} \quad x = b \div a.$$

For example, we can express the fact that $7 \cdot 8 = 56$ by writing $8 = 56/7$ or $7 = 56/8$.

In case $a = 7$ and $b = 1$, there is no integer x such that $ax = b$; in other words there is no integer x such that $7 x = 1$. You may ask: "Is there a number x such that $7 x = 1$?"

[1] Cf. reference **31**, Chap. III.

If by "number" you mean integer, then the answer is "no." What is now required is the inventive genius of a daring human who will rise to the occasion and say "Let there be a number x such that $7x = 1$, let the name of this x be one-seventh, and let x be denoted by the symbol $1/7$." We accept this creation, and likewise create enough other numbers x so that

$$ax = b$$

will have a solution if a and b are integers and $a \neq 0$. Moreover, we understand that $x = b/a$ when and only when $ax = b$ and $a \neq 0$. This is in accord with our definition of the symbol b/a when b is divisible by a.

The numbers x which we have just defined are "quotients of integers" or "ratios of integers," and are called **rational numbers.** These numbers are called **rational** (ratio-nal) because they are *ratios* of integers; not because they have attributes opposite to those of inmates of asylums.

It is opportune here to point out that there are limitations on what numbers we can create. Even an omnipotent Diety cannot create a black rock which is entirely white because being white would be incompatible with the property of being black. So man "made in God's image" cannot create a *number* x such that $0 \cdot x = b$ for $b \neq 0$ since the existence of such a number would be incompatible with our conception of a number. The incompatibility between the existence of such a number and our ideas about numbers can be shown in several ways. One way is to show as follows that multiplication by such a "number" would not be associative. Suppose $0 \cdot x = b$ for some number x when b is an integer. Then, since $0 \cdot x$ and b are the same number, we can multiply them by 0 and have

$$0(0 \cdot x) = 0 \cdot b.$$

But, by the associative law, the left side is equal to $(0 \cdot 0)x = 0 \cdot x = b$ whereas, since b is an integer, we know that the right side, namely, $0 \cdot b$ is zero. This shows that if the associative law is to hold and $0 \cdot x = b$ for some number x, then b must be 0. But if $b = 0$, any number x will do. In other words, the quotient $b/0$ cannot be a number (that is, cannot have all the properties of a number) unless b is 0; in that case $0/0$ can be given any value you please. Since we wish the quotient b/a not only to have the properties of numbers but to have just one value, **we exclude $a = 0$ as a divisor.** We assume, though it is not hard to prove it, that there is only one number x such that $ax = b$ when $a \neq 0$.

Of course, it might conceivably be true that the numbers $x = b/a$ which we have created would violate some of the laws associated with numbers, but it can be shown that such numbers can be made to obey all the laws we want numbers to obey. We shall for the most part take this for granted.

It should be noticed that integers are special rational numbers. For example, 7 can be written in such forms as $7/1, 14/2$, and $(-56)/(-8)$. If n is an integer, then $n \cdot 0 = 0$; hence 0 can be written as $0/1$ or $0/6$ or $0/(-8)$.

The symbols $7/1, 3/5$, etc., are also called **fractions.** In fact, any number

$$a/b \quad \text{or} \quad \frac{a}{b}$$

where a and b are numbers is called a **fraction.** As you know, a is called the **numerator** and b the **denominator** of the fraction. From the definition of a rational number it is clear that all rational numbers can be expressed as fractions whose numerator is an integer and whose denominator is an integer different from zero. However, not all fractions are rational numbers. When later on we prove that $\sqrt{2}$ is not a rational number we shall know that $\sqrt{2}/3$ is not a rational number though it is a fraction. Rational fractions used to be called **common fractions** or, much earlier, **vulgar fractions.**

It should be pointed out that what we have described above is in exact conformity with, shall we say, our notion of one-seventh of a pie; for when we visualize one-seventh of a pie we think of that portion of the pie which, when increased sevenfold, is the whole pie. This is exactly equivalent to saying $7x = 1$ where 1 stands for the whole pie and x stands for each portion. But this point of view makes the existence of such a number no less a matter of invention. Any one of us would have considerable difficulty in dividing a pie into seven equal parts, but this would not make us hesitate for a moment to speak about a seventh of a pie.

When, however, we speak of fractions whose numerators are not 1 we have a slightly different story. By our above definition $3/7 = x$ means $7x = 3$ which, in the language of the previous paragraph, is connected with dividing 3 into 7 equal parts. Dividing a set of 3 pies into 7 equal parts is quite different from dividing one pie into 7 equal parts and taking 3 of these parts, even though the quantity of pie in $3/7$ is the same, assuming that the pies are of the same size and none comes off on the knife. Since the second concept has some advantages let us see how we should show the two concepts to be the same quantitatively. Let $x = 1/7$. Then we know $7x = 1$. If we assume that rational numbers have the same associative and commutative properties as do integers, the last equation implies $7(3x) = 3$, and by definition $3x = 3/7$; that is, $3(1/7) = 3/7$. The proof that this holds for all rational numbers is given as an exercise below.

6. Addition and multiplication.

Without attempting further to justify the assumptions, we in this book assume that rational numbers can be added and multiplied and that for any rational numbers a, b, and c we have the following properties:

$$a + b = b + a \qquad\qquad ab = ba$$
$$(a + b) + c = a + (b + c) \qquad (ab)c = a(bc)$$
$$a + 0 = a \qquad\qquad a \cdot 0 = 0$$
$$a(b + c) = ab + ac$$

For rational numbers a and b we write $x = b - a$ when $a + x = b$, and $x = b/a$ when $a \neq 0$ and $ax = b$.

These properties can be used to obtain familiar formulas involving fractions. In the first place, everyone should believe that

$$\frac{ka}{kb} = \frac{a}{b} \quad \text{if} \quad kb \neq 0.$$

This is easily shown by letting $x = a/b$ which means $bx = a$. Since the number on the left is equal to that on the right, we may multiply [1] both sides of the equation by k and have $k(bx) = ka$. By the associative law $(kb)x = ka$ which means that x is not only equal to a/b but to ka/kb as well.

Everyone should believe that

$$\tfrac{3}{17} + \tfrac{5}{17} = \tfrac{8}{17}.$$

Using the result of Exercise 6 below and the distributive law for rational numbers we can establish this result as follows:

$$\tfrac{3}{17} + \tfrac{5}{17} = 3(\tfrac{1}{17}) + 5(\tfrac{1}{17}) = (3 + 5)(\tfrac{1}{17})$$
$$= 8(\tfrac{1}{17}) = \tfrac{8}{17}.$$

Notice that we are really *defining* 8/17 to be the sum of 3/17 and 5/17 in order that the distributive law will hold, or, less technically, so that seventeenths will have one of the same properties that dogs have: e.g., that three dogs plus five dogs is eight dogs. The general result of which this is an example is

$$\frac{b}{a} + \frac{c}{a} = \frac{(b + c)}{a}.$$

On the other hand, no one should believe that $17/3 + 17/5 = 17/8$, but alas, some do! To find its true value, we could begin as above:

$$\tfrac{17}{3} + \tfrac{17}{5} = 17(\tfrac{1}{3}) + 17(\tfrac{1}{5}) = 17(\tfrac{1}{3} + \tfrac{1}{5}).$$

[1] Notice that we here use an important property of equality, namely, that $r = t$ implies $wr = wt$, no matter what w is.

But what is $1/3 + 1/5$? Surely it is not $1/8$ for this is less than $1/3$ as well as $1/5$. Here a familiar trick is used. We have seen that we can easily add two fractions whose denominators are equal. Hence we replace $1/3$ by its equal, $5/15$; and $1/5$ by its equal, $3/15$. Then

$$\tfrac{1}{3} + \tfrac{1}{5} = \tfrac{5}{15} + \tfrac{3}{15} = \tfrac{8}{15}$$

and $\quad \tfrac{17}{3} + \tfrac{17}{5} = 17(\tfrac{8}{15}) = \tfrac{(17 \cdot 8)}{15} = \tfrac{136}{15}.$

We have, of course, here used a more roundabout method than is necessary. We could more expeditiously have used the trick on the original sum and would have the familiar

$$\tfrac{17}{3} + \tfrac{17}{5} = \tfrac{85}{15} + \tfrac{51}{15} = \tfrac{136}{15}.$$

In general

$$\frac{a}{b} + \frac{c}{d} = \frac{ad}{bd} + \frac{cb}{bd} = \frac{(ad + cb)}{bd}.$$

We now look at some products of fractions. It is true that

$$\tfrac{2}{7} \cdot \tfrac{3}{5} = \tfrac{6}{35}.$$

Let us see how we can use the rules to obtain this result. Put $a = 2/7$, $b = 3/5$. Then $7\,a = 2$ and $5\,b = 3$ and therefore $(7\,a)(5\,b) = 2 \cdot 3$, $35\,ab = 6$ and $ab = 6/35$.

EXERCISES

1. Find the value of each of the following:

$$\frac{5}{6} + \frac{3 - 4}{7}, \quad -\frac{7 - 2}{7} + \frac{6}{11}.$$

2. Which of the following equalities hold whatever numbers a and b are, provided no denominator is zero? Give the reasons for your answer.

$a.\ \dfrac{3\,a + b}{3\,c} = \dfrac{a + b}{c}.$

$b.\ \dfrac{7\,a}{7\,b + c} = \dfrac{a}{b + c}.$

$c.\ \dfrac{5\,a + 5\,b}{5\,c + 5\,d} = \dfrac{a + b}{c + d}.$

3. Using the results of Exercises 6 and 7 where necessary, find the values of the following:

a. $\frac{3}{7} \cdot \frac{4}{9} \cdot \frac{7}{8}$.

b. $\frac{3}{7}(\frac{4}{9} + \frac{7}{2})$.

c. $\dfrac{3+5}{6} \cdot \dfrac{4}{7}$.

d. $\frac{3}{8} \cdot \frac{5}{9} + \frac{1}{6} \cdot \frac{5}{16}$.

e. $\frac{3}{4}(\frac{3}{2} - \frac{4}{7})$.

4. Express as a single fraction:

a. $\dfrac{a}{b} + \dfrac{c}{d}$.

b. $\dfrac{a}{b} - \dfrac{c}{d}$.

c. $\dfrac{a}{b} + \dfrac{c}{d} + \dfrac{e}{f}$.

d. $a\left(\dfrac{b}{c} - \dfrac{d}{e}\right)$.

e. $\dfrac{a}{b} - \dfrac{c}{d}\left(\dfrac{e}{f} - \dfrac{g}{h}\right)$.

5. Prove that if a and b are any two rational numbers and $ab = 0$, then $a = 0$ or $b = 0$.

6. Prove that for all integers a and b with $b \neq 0$, $a(1/b) = a/b$.

7. Prove that

$$\frac{a}{b} \cdot \frac{c}{d} = \frac{ac}{bd}.$$

8. Prove that for no integers b and c, neither being 0, is it true that $1/c + 1/b = 1/(c + b)$.

9. Show that $-(a/b) = (-a)/b = a/(-b)$ if $b \neq 0$.

7. Division.

Is there a rational number x such that $ax = b$? The answer is "It depends." If a and b are *both* zero, the equation is satisfied by *every* rational number. We showed in section 5 that if a is zero and b is not zero, the equation cannot be satisfied by any number which has the usual properties of numbers of our acquaintance. Thus, if $a = 0$, then the equation $ax = b$ *never* has a unique solution for x: the equation has either many solutions (when $b = 0$) or none whatever (when $b \neq 0$).

In case a and b are rational and $a \neq 0$, we can show that the equation

$$ax = b$$

has exactly one solution. If

$$a = \frac{j}{k}, \quad b = \frac{m}{n},$$

where j, k, m, and n are integers and none, except perhaps m, are zero, then

$$a \cdot \frac{k}{j} \cdot \frac{m}{n} = \frac{j}{k} \cdot \frac{k}{j} \cdot \frac{m}{n} = \frac{m}{n} = b.$$

Therefore, the equation is satisfied when

$$x = \left(\frac{k}{j}\right)\left(\frac{m}{n}\right) = \frac{km}{jn}.$$

This means: *to divide b by a, multiply b by the reciprocal of a.*

This shows that there is a rational number x such that $ax = b$ when a and b are rational and $a \neq 0$. That x is the only rational number that satisfies the equation can be shown as follows: let $ax = b$ and $ay = b$. This implies that $ax = ay$. Since a is not zero, we can find a rational number r such that $ra = 1$ (that is, $r = 1/a$). Multiply both sides of $ax = ay$ by r and get $rax = ray$, that is, $x = y$.

EXERCISES

1. Find the values of

a. $\dfrac{\frac{16}{8}}{4}$. *Ans.* $\frac{1}{2}$.

b. $\dfrac{16}{\frac{8}{4}}$.

c. $\dfrac{5(\frac{1}{2} - \frac{3}{4})}{\frac{3}{7}}$. *Ans.* $\dfrac{-35}{12}$.

d. $\dfrac{7}{8} + \dfrac{2(\frac{1}{3} - \frac{1}{4})}{\frac{8}{3}}$.

e. $\dfrac{3(\frac{4}{3} - \frac{1}{5})}{3(\frac{1}{2} - 1)}$.

f. $\dfrac{\frac{a}{b}}{\frac{c}{d}}$.

g. $\dfrac{\frac{x(a - b)}{c} + \frac{y}{d}}{\frac{x}{cd}}$. *Ans.* $\dfrac{xad - xbd + yc}{x}$.

h. $\dfrac{\frac{r}{s} - \frac{t(x - y)}{w}}{\frac{x}{s} - \frac{y}{w}}$.

2. How many possible different meanings can the following ambiguous symbols have?

a. 1/2/3.

b. $x/y/z$.

c. 1/2/3/4.

d. $t/x/y/z$.

3. If a and b are rational numbers, prove that ab is a rational number.

4. Prove that if a and b are rational numbers and $a \neq 0$, then b/a is a rational number.

8. Decimals.

Since we use the decimal system of reckoning (rather than the dozal or some other system) it is as natural to represent rational numbers as sums of hundreds, tens, units, tenths, hundredths, etc., as it is to represent positive integers as sums of units, tens, hundreds, etc. Fractions such as 3/10, 271/100, and 52/1000 in which the denominator is 10, 100, 1000, etc., occur so frequently that a special way of writing them has been invented. Just as $20 = 2(10)$, $200 = 2(100)$, $2000 = 2(1000)$, we have $2 = 2(1)$, $.2 = 2(1/10)$, $.02 = 2(1/100)$, $.002 = 2(1/1000)$. Every time we divide by 10 we move the *decimal point* one place to the left and every time we multiply by 10 we move it one place to the right, adding zeros if necessary to indicate its position.

The monetary system of the United States makes use of decimals; for example, 712 cents make 7.12 dollars; and 2.99 dollars make 299 cents. The metric system of measurement was devised to fit in with our decimal notation: 100 centimeters = 1 meter, 1 decimeter = .1 of a meter, etc. On the other hand, the division of a foot into twelve inches is not easily adaptable to this system, for 1 inch = .08333 ⋯ of a foot. It would, however, fit in nicely with the dozal system, for in that case .1 would mean one "dozenth" (that is, one twelfth) and we should have 1 inch = .1 of a foot in dozals while 1 decimeter = .124972 ⋯ of a meter in the same system.

We have all had experience with multiplication and division of decimals but it may not be without profit here briefly to explore the subject in order to find the reasons back of familiar rules. (Sometimes, alas, they are not as familiar as they should be.) We have the following

Rule: In multiplying two decimals, the number of places to the right of the decimal point in the product is the sum of the number of places to the right of the decimal point in the numbers of the product.

The following illustration will show the reason back of the rule; suppose we multiply 1.576 by 2.32. The former is $1576(1/10^3)$ and the latter is $232(1/10^2)$. Hence the product is equal to

$$232(1576)(1/10^3)(1/10^2).$$

But

$$\frac{1}{10^3} \cdot \frac{1}{10^2} = \frac{1}{10^3 \cdot 10^2} = \frac{1}{10^5}$$

since 10 taken as a product three times multiplied by 10 times 10, is 10 taken as a product five times. Thus, to get the product of the given numbers, we multiply 232 by 1576 and place the decimal point five places from the right. If our numbers were .001576 and .0232 respectively we would have $1/10^6$ instead of $1/10^3$, $1/10^4$ instead of $1/10^2$, and hence $1/10^{10}$ instead of $1/10^5$. Notice, by the way, that the number of digits to the right of the decimal point of any number (not counting terminal zeros) is equal to the least power of 10 by which the number need be multiplied to become an integer.

The rule for division is just the reverse of that for multiplication. Since the number of places to the right of the decimal point in the product is the sum of the number of places for the members of the product, so the number of places in the quotient is the number of places in the dividend minus the number of places in the divisor. For instance, in dividing 5.37 by 253.2 we write

$$\begin{array}{r} .02 \\ 253.2\,\overline{\smash)\,5.370} \\ 5.064 \end{array}$$

The position of the decimal point in 5.064 is determined by that in 5.37, the dividend. Since 5.064 has three places to the right of the decimal point and 253.2 has one, the quotient must have $3 - 1 = 2$ places to the right of the decimal point.

Some numbers are easily expressed (or represented) in decimal form. For example, $2/5 = .4$, $-10/4 = -2.5$, $3/8 = .375$. But there are other numbers such as $1/3$, $5/7$, and $1492/1776$ for which representation in decimal form is not so simple. We have all written

$$\frac{.3333333 \cdots}{3 \,\overline{\big)\, 1.0000000 \cdots}}$$

and then

$$\tfrac{1}{3} = .3333333 \cdots.$$

To examine the connection between the number $1/3$ and the **nonterminating decimal** $.3333333 \cdots$, let

$$s_1 = .3$$
$$s_2 = .33$$
$$s_3 = .333$$
$$s_4 = .3333$$

. . .

Each of the numbers[1] s_1, s_2, s_3, $\cdots$ is a "terminating decimal," that is, we can write the entire decimal — there are no dots in any single s indicating any desire to go on like this forever. It is instructive to compute the differences $1/3 - s_1$, $1/3 - s_2$, $1/3 - s_3$, $\cdots$. We find that

$$\frac{1}{3} - s_1 = \frac{1}{3} - \frac{3}{10} = \frac{10-9}{30} = \frac{1}{30} = \frac{1}{3(10)},$$

$$\frac{1}{3} - s_2 = \frac{1}{3} - \frac{33}{100} = \frac{1}{300} = \frac{1}{3(10^2)},$$

$$\frac{1}{3} - s_3 = \frac{1}{3} - \frac{333}{1000} = \frac{1}{3000} = \frac{1}{3(10^3)},$$

and, in general,

$$\frac{1}{3} - s_n = \frac{1}{3(10^n)}$$

where 10^n is 10 taken as a product n times. It thus appears that, when we take more and more terms of the decimal $.33333 \cdots$ to have successively $.3$, $.33$, $.333$, $\cdots$, we get num-

---

[1] We read s_1 "s sub-one," etc.

bers nearer and nearer to 1/3. Furthermore, no matter how small a number you name I can carry out the division to enough places so that the terminating decimal s_n differs from 1/3 by less than that number. This can be done since n can be taken large enough so that $\dfrac{1}{3(10^n)}$ is smaller than the number you named. This fact is conveniently expressed by the simple statement "the sequence .3, .33, .333, $\cdots$ converges [1] to 1/3." We say that ".3333 $\cdots$ is the **decimal expansion** of 1/3" or that ".3333 $\cdots$ **converges** to 1/3."

Division of 3.0000 $\cdots$ by 7 gives .428571428571 $\cdots$; and it can be shown that

$$\tfrac{3}{7} = .428571428571 \cdots$$

in the sense that the sequence .4, .42, .428, .4285, $\cdots$ converges to 3/7. It can also be shown that

$$\tfrac{15052}{3300} = 4.561212121212 \cdots$$

where the block 12 keeps repeating without end; for this reason the decimal is called a **repeating decimal** (even though there are some terms at the beginning before the repetition starts). Terminating decimals can be made into nonterminating decimals by adding zeros; for example,

$$\tfrac{1}{4} = .25 = .25000000 \cdots$$

where the 0 keeps repeating. Thus the decimal value of 1/4 can be considered to be a repeating decimal.

The expression

$$.101001000100001000001 \cdots,$$

in which each bunch of zeros contains one more than the preceding bunch, is not a repeating decimal; hence it cannot be the decimal expansion of a rational number.

---

[1] In more advanced courses it is necessary to have a precise definition of convergence. A sequence s_n is said to converge to s if, corresponding to each positive number ϵ, there is a number N such that the numerical (absolute) value of $s - s_n$ is less than ϵ when n is greater than N.

Exercise: If $s_1 = .3$, $s_2 = .33$, $s_3 = .333$, $\cdots$ and $s = 1/3$, find a number N which corresponds to the value 1/1,000,000 of ϵ.

We shall show later on that every repeating decimal converges to a rational number. Combining this result with the converse statement given in Exercise 18 below we see that *the rational numbers are precisely (that is, all and only) those numbers which can be expressed as repeating decimals or terminating decimals.*

EXERCISES

1. Five gross, four doza-two inches is how many feet? How would you write your answer in dozals?

2. Six hundred twenty-five eggs is how many gross in dozals?

3. In the binary scale suppose a quart is represented by the number 1. How would you represent the two quantities, a pint and a gill, in what, for the binary scale, corresponds to decimals for the decimal scale of notation?

4. Divide 1.0626 by 253 and carefully explain the reasons for your results and methods.

5. Show that the difference between 3/7 and .428571 is 3/7,000,000.

6. Show that .428 is less than 3/7 and that .429 is greater than 3/7.

7. Find a terminating decimal whose value differs from 1/11 by less than 1/1,000,000.

8. Does the addition of the decimal expansions of 1/7 and 1/3 give the decimal expansion of 10/21?

9. The first six numbers of the decimal form of 1/7 are 142857. 3(142857) = 428571; 2(142857) = 285714. Can you give any explanation of this phenomenon? [1] What multiple of 142857 is 857142? Does multiplication of 142857 by 8 fit into the scheme?

10. Find the repeating decimal for 1/101.

11. Find a rational number whose value is between 1/6 and 5/31.

12. Find a rational number whose value is between two rational numbers a and b.

13. Which is greater: 4/11 or 3/8? 23/11 or 27/13?

14. Two $23\frac{1}{2}$-ounce packages of soap can be bought for 37 cents or one 4-pound 5-ounce package for 55 cents. If the weights are net weights, which is the more economical purchase?

15. Brand A of ginger ale sells at three bottles for 25 cents, and brand B three bottles for 20 cents. If each bottle of the former holds 16 ounces and each of the latter 13 ounces, which is the cheaper?

* 16. What will be the integers b for which $1/b$ can be expressed as a terminating decimal?

* 17. Obtain the decimal form of 1/7 and find whether you can multiply it by 3 to get the decimal form of 3/7.

[1] See references **4** and **20**.

* 18. Show that the nonterminating decimal expansion of each rational number is a repeating decimal.

* 19. Show that the part of the decimal value of $1/m$ which repeats has less than m digits if m is a positive integer; e.g., $1/7$ has less than 7 digits in the repeating part of its decimal expansion.

** 20. Show that if p is a prime number, the number of digits in the repeating part of the decimal value of $1/p$ is a divisor of $p - 1$.

** 21. Show that the nonterminating decimal expansion of each rational number converges to the number.

9. Squares.

The squares of the positive integers 1, 2, 3, 4, $\cdots$ are 1, 4, 9, 16, $\cdots$; the square of 0 is 0; and the squares of the negative integers -1, -2, -3, -4, $\cdots$ are 1, 4, 9, 16, $\cdots$. Thus it is apparent that there is no integer whose square is 2.

We are now going to prove that *there is no rational number whose square is* 2. We do this by assuming that r is a rational number whose square is 2 and then showing that the assumption is absurd. Let $r = p/q$ where p and q are integers having no common integral factor greater than 1. (Each rational number, and in particular r, can be represented in this way.) The assumption that $r^2 = 2$ means that $(p/q)^2 = 2$ and hence that $p^2/q^2 = 2$ and

$$p^2 = 2\,q^2.$$

Since 2 is a factor of $2\,q^2$ and $p^2 = 2\,q^2$, 2 must be a factor[1] of p^2. Therefore, p is divisible by 2, for if p were an odd number then p^2 would also be odd. Since p is divisible by 2, we can write $p = 2\,t$ where t is an integer. Then $p^2 = 2\,q^2$ implies that $4\,t^2 = 2\,q^2$ and hence $2\,t^2 = q^2$. Thus, q^2 must be divisible by 2. Hence q is divisible by 2. We have shown that 2 is a common factor of p and q; this contradicts the fact that p and q have no common positive integral factor greater than 1. Therefore, the assumption that there is a rational number whose square is 2 is absurd.

---

[1] This follows from the fact noted in Chap. II that any number, in particular p^2, has only one factorization into prime factors aside from the order of the factors.

EXERCISES

1. Prove that there is no rational number whose square is 3.
2. Prove that there is no rational number whose cube is 2.

10. Real numbers.

The mere fact that there is no rational number whose square is 2 indicates that the set of rational numbers is not adequate to give a solution of $x^2 = 2$, and that a larger class of numbers should be created. The simplest thing for us to do is to take advantage of the inventive genius of the first man who said "Let there be enough numbers so that there is one for each nonterminating or terminating decimal, and let these numbers be called **real numbers.**"

In more advanced courses, properties of addition and multiplication of these real numbers are given; these properties include the commutative, distributive, and other properties which apply to integers and rational numbers.[1]

It can be seen that our definition of a real number implies that there is, in a certain sense, a line segment whose length is any given real number. To this end consider the non-repeating decimal we previously mentioned

$$.101001000100001 \cdots .$$

We can draw a line segment of length .1, then one of length .101, then one of length .101001 and so on. Each line segment will differ in length from the preceding by an ever smaller amount. We can *imagine* a line segment to which the process approximates more and more closely and associate this "imaginary" line segment with the nonterminating decimal. This process can be carried through for any real number. Also, if a and b are any two real numbers whose nonterminating decimals we can find, we can determine which is the greater. We then say that the real numbers have an **order** or are **ordered.**

We have stated and partially proved the fact that the real numbers which correspond to repeating decimals are

---

[1] See reference **11**, Chap. IV, for the concept called "The Dedekind cut."

rational numbers; those real numbers which correspond to nonrepeating decimals are not rational numbers and are called **irrational numbers.** All irrational numbers are real. It can be proved that there is a positive real number whose square is 2; this number is denoted by $\sqrt{2}$ (or $2^{\frac{1}{2}}$). The proof lies beyond the scope of this book but one of the things on which it rests is that the ordinary process for extraction of square roots of numbers leads to a nonterminating decimal which represents or converges to $\sqrt{2}$; that is, the sequence

$$1.4^2, \ 1.41^2, \ 1.414^2, \ 1.4142^2, \ \cdots$$

converges to 2. On the basis of this assumption that $\sqrt{2}$ is indeed a real number and using our proof that it is not rational we can say that $\sqrt{2}$ is an irrational number.

The irrationality of $\sqrt{2}$ is equivalent to the geometrical statement that the length of the diagonal of a square is incommensurable with its sides. This means, as you recall, that there is no unit of length such that the length of the diagonal as well as the length of the side is an integral multiple of such a unit; for supposing the diagonal were p units long and the side q units long, then the diagonal would be p/q times the length of the side, that is, a rational multiple of the side. This we have proved to be false. It was in this geometrical form that Euclid and others before him proved the irrationality of $\sqrt{2}$.

The irrational numbers which we have mentioned above are all roots of equations with "rational coefficients" like $x^2 - 2 = 0$, which has the root $\sqrt{2}$, or $x^2 - 3 = 0$, which has the root $\sqrt{3}$. (The **coefficients** are the multipliers of the powers of x; 1, 0, and -2 in the first equation and 1, 0, and -3 in the second equation.) In fact, $x^3 + ax^2 + bx + c = 0$ has irrational roots for many rational numbers a, b, and c; one can even go further and say that, if $c = 1$ and a and b are integers, none of its roots are rational unless $a = b$ or $a + 2 = -b$. The number π is, however, an altogether "different kind" of irrational number for it is not a root of any such equation. (Recall that π is defined to be the ratio of the circumference of a circle to its diameter.) This statement is made more precise when we say (1) that any root of an equation

$$ax^n + bx^{n-1} + \cdots + q = 0,$$

where the coefficients a, b, $\cdots q$ are rational numbers, is called an **algebraic number;** (2) it is true that π is not an algebraic number. This second statement has been proved but it is too difficult to be included here. Any number, π included, which is not an algebraic number is called a **transcendental number.** Those who have had some trigonometry will be interested to know that the trigonometric functions of many angles are transcendental numbers. All rational numbers are algebraic numbers. Some irrational numbers are algebraic and some are not.

Of course, π being transcendental is all the more irrational. Its decimal is not a repeating decimal as a consequence. It may be of interest to know that its value to thirty decimal places is

3.14159 26535 89793 23846 26433 83280.

William Shanks, about 1853, computed π to 707 decimal places. We hope the computation did not lead him to an early grave.

EXERCISES

1. Show how to draw a line $\sqrt{3}$ inches long.

2. Prove that 2 lies between the squares of the numbers 1.4142 and 1.4143.

3. Find a rational number which is closer to $\sqrt[3]{3}$ than .1; that is, such that the difference between $\sqrt[3]{3}$ and the number you find is to be less than .1.

4. Prove that the sum of a rational number and an irrational number is an irrational number. Is the result the same if "sum" is replaced by "product"?

5. Can the sum of two irrational numbers ever be rational?

6. If $a + bx = 0$, where a and b are rational but x is irrational, prove that a and b are both zero.

* 7. Draw a line $\sqrt{2}$ inches long.

*8. Show that between any two irrational numbers there is always a rational number.

*9. Show that between any two rational numbers there is always an irrational number.

11. Complex numbers.

We "complete" (see the end of the section) our process of obtaining more and more extensive number systems by making the step to "imaginary" numbers and "complex" numbers.

The real numbers may be divided (as the integers were) into three classes: positive real numbers, zero, and negative real numbers. Since the square of each positive real number is positive, the square of 0 is 0, and the square of each negative real number is positive, it follows that there is no real number x such that $x^2 = -1$. It is now appropriate to ask "Is there a number whose square is -1?" If by number one means "real number," the answer is "no." Anyone who has really understood the steps in the discussion of negative integers, of rational numbers, and of real numbers should rise immediately and proclaim "Let there be a number whose square is -1." This new number is called i, the initial letter of "imaginary." The number i is called **imaginary**. Of course, this does not mean that i is imaginary in the sense that there really is no such number; we "imagine" a number i whose square is -1 in exactly the same way that we "imagine" a number whose square is 2. Anyone skilled in mathematics or in one of the numerous sciences which require use of mathematics knows that the number i plays an important role in the world of affairs. As an actual matter of fact, the usefulness of the number i is not impaired by its unfortunate name which naturally tempts the young and gullible to think that it is somehow far more mysterious than the "real numbers."

Numbers of the form

$$z = x + iy$$

where x and y are real numbers and $i^2 = -1$ are called **complex numbers.** The laws governing the multiplication, division, addition, and subtraction of complex numbers are the same as those of real or rational numbers, but we do not attempt to prove it here. Notice, however, that non-real numbers cannot be represented as points on a line as the real numbers could be. Hence not without certain extensions of our ideas, which we have not time to make here, can we speak of one complex number being greater than another, at least in the same sense that real numbers are. In other words, the complex numbers are not ordered, at least in the same way as are the real numbers.[1]

At this point we should call attention to the distinction between "complex numbers" and "imaginary numbers": $z = x + iy$ is a complex number no matter what real numbers x and y are, but it is **imaginary** only if $y \neq 0$. In other words, all real numbers are complex numbers but none are imaginary numbers. Each complex number is either real or imaginary but not both.

Those who may have been curious about the word "complete" in the first sentence of this section will perhaps be enlightened by the following remarks. We said, first, that real numbers and, second, complex numbers have the same properties with regard to multiplication, etc., as do the rational numbers. When we say that our process of finding numbers is "complete" we mean that there are no more; that is, that all quantities which behave like (that is, have the same properties as) rational numbers as far as addition, subtraction, multiplication, and division are concerned are essentially complex numbers. [This statement is another thing we do not prove here.] In more advanced mathematics we sometimes speak of things not complex numbers which we may call **numbers** (some of them are sometimes called **hypercomplex** numbers and might be called **super-**

---

[1] See reference **25**, pp. 100 ff., for a method of finding a point which represents a complex number.

complex numbers in accordance with the advertising of the day) but in one or more respects they misbehave. About the most widely used misbehaving "numbers" are some which act like this:

$$ab = -ba.$$

The physicists like them and so do many mathematicians but we, being good little boys, will not associate with them.

The relationships between the various kinds of numbers we have defined in this book are embodied in the following diagram:

All complex numbers are $\begin{cases} \text{real or imaginary} \\ \text{algebraic or transcendental} \end{cases}$

All real numbers are $\begin{cases} \text{positive, negative, or zero} \\ \text{rational or irrational} \end{cases}$

The diagram may also be read backwards, e.g., all rational and irrational numbers are real. Furthermore, no two names on any one line can truthfully be used to describe one number, e.g., a number cannot be both real and imaginary.

EXERCISES

1. Using distributive and associative laws and the fact that $i^2 = -1$, prove that

$$(1 + i)(2 + i)(3 + i) = 10\,i.$$

2. Simplify the fraction

$$\frac{3 + i}{2 - i}$$

by multiplying the numerator and denominator by $2 + i$ and simplifying the result.

3. Show (by substituting $2 + 3\,i$ for x in the left side) that the equation

$$x^2 - 4\,x + 13 = 0$$

is satisfied when $x = 2 + 3\,i$.

4. Show that

$$x^2 - 4\,x + 13 = (x - 2)^2 + 9$$

and then explain why there is no real value of x for which

$$x^2 - 4\,x + 13 = 0.$$

5. Are any complex numbers rational? Are any imaginary numbers rational? If in either case your answer is "yes" give an example.

6. Ascribe all possible names to each of the following numbers:

     a. $\sqrt{2}$.   *Ans.* complex, real, algebraic, positive, irrational.
     b. $-3\sqrt{2}/\sqrt{2}$.   *c.* $\sqrt[3]{-2}$.   *d.* $\sqrt{-5}$.
     e. .123123 ⋯.   *f.* .121121112 ⋯.   *g.* π.

7. Following the pattern of Chapter I, draw circles indicating the relationships among the types of numbers we have so far considered in this book.

* 8. One might define a complex number $a + bi$ to be "greater than" $c + di$ if $a > c$, or $a = c$ and $b > d$. Which of the following properties hold for *every* set of complex numbers p, q, r, s?

     a. $p > q$ and $q > r$ implies $p > r$,
     b. $p > q$ and $r > s$ implies $p + r > q + s$,
     c. $p > 0$ and $r > s$ implies $pr > ps$.

In each case either prove your answer to be "yes" or give an example in which the implication does not hold.

12. Topics for further study.

    1. Irrational numbers: reference **11**, Chap. 4.

    2. Approximations to π: references **11**, p. 68; **25**, pp. 72–79.

    3. Revolving or cyclic numbers: references **4; 20; 28**, Chap. 5.

CHAPTER IV

Algebra

Myself when young did eagerly frequent
Doctor and Saint, and heard great argument
   About it and about: but evermore
Came out by the same door wherein I went.

. . . . . .

Waste not your Hour, nor in the vain pursuit
Of This and That endeavour and dispute;
   Better be jocund with the fruitful Grape
Than sadden after none, or bitter, Fruit.

. . . . . .

Ah, but my Computations, People say,
Reduced the Year to better reckoning? — Nay,
   'T was only striking from the Calendar
Unborn To-morrow, and dead Yesterday.

1. Introduction.

Omar Khayyám, so the story goes, was one of the three favorite pupils of Imám Mowaffak of Naishapur, one of the greatest of the wise men of Khorassan. These three pupils vowed that the one of them to whom the greatest fortune fell would share it equally with the rest. When one of them became Vizier to the Sultan he made Omar Khayyám official astronomer in the court of the Sultan. And this man, of whom we are apt to think only as the author of the *Rubáiyát*, became one of the greatest mathematicians of his time. Among other things, he revised the Arabian calendar and wrote a book on algebra which David Eugene Smith declares "was the best that the Persian writers produced." In this book are the first recorded instances of certain results that we use today. It is rather interesting that in the fourteenth edition of the *Encyclopædia Brittanica* the only reference in the index to this Persian poet is to a quarter-page discussion of some of his remarkable results in algebra. This fact is probably in line with Omar Khayyám's opinion of the relative values of his achievements in life, all his talk notwithstanding.

But Omar Khayyám was not by any means the first algebraist. Ahmes, the Egyptian, used what might be called **algebra** about 1550 B.C. and Diophantus, the Greek, wrote the first treatise on algebra about 275 A.D. But who the early algebraists were cannot be decided upon until we agree on what we mean by algebra and on this point there' seems to be considerable difference of opinion. However, even though the advanced mathematician might not agree, most of those who read these pages would probably be willing to say that algebra consists fundamentally in allowing letters to stand for numbers, and either considering such a representation as a kind of formula for a general result or the basis for a general process which one could carry through with any set of numbers. Of course, symbolism of this kind came historically some time after the solution of problems that we should today solve by means of algebra.

Leaving this very brief discussion of the history of algebra to glance back over the first chapters of this book we see that this is not the first time this symbolism, which we may call **algebra,** occurs. We have already used letters in place of numbers to gain the fundamental advantage of not saying exactly what we are talking about. To say that all men have noses (a statement which is true in the vast majority of cases) is a much stronger statement than saying that John has a nose. From the general statement I can conclude that John has a nose or that Jack has a nose or that Edward has a nose — in general, "all God's children have a nose." But to know merely that John has a nose is no help whatever in showing that Edward has one also, unless perhaps one uses John to see what a thing called a nose looks like. Furthermore, to say that a man traveling 20 miles an hour for 5 hours will have gone $5 \cdot 20 = 100$ miles expresses an isolated result but to say that $T \cdot R = D$, when R is the constant rate in miles per hour and T is the number of hours, gives the distance, D, in miles, is to state a result for every rate and every period of time. It is the essence of algebra, as we know it, that by writing letters in place of numbers we can obtain a result that applies to a large set of numbers. The usual algebra taught in high school tends to leave the impression that its chief use is in solving equations. It is true that later on in this chapter we shall consider this topic, but such a use of algebra is minor compared to its use in proving certain general results and in deriving useful formulas.

2. Square numbers.

By way of illustrating the use of algebra in arriving at certain general conclusions we first consider an interesting property of the list of squares of numbers. Recall first that 3 squared, which is written 3^2, is another way of writing $3 \cdot 3$. The list of the squares of the integers then, will begin like this:

1 4 9 16 25 36 49 64 81 100 121 144 169 196 225.

(We stop when we become tired.) Compiling the list becomes more and more laborious as we proceed. Unless we are very good at mental arithmetic, we cannot, without using paper and pencil, continue the list beyond what we have memorized; that is, we cannot unless we notice something about this list. There is, on investigation, a certain property of the differences of the successive terms of the sequence. We notice that $4 - 1 = 3$, $9 - 4 = 5$, $16 - 9 = 7$, $\cdots$ and continuing we see that the successive differences are these:

3 5 7 9 11 13 15 17 19 21 23 25 27 29.

That looks as if the next square beyond 225 is $225 + 31 = 256$. And, sure enough, multiplication shows that $16^2 = 256$! We

can continue, to get $17^2 = 256 + 33 = 289$ since by now we
are rather confident that the system will work.

We may consider ourselves clever at having discovered
the above system which *seems* to hold. If so, the wind will
be taken out of our sails by the following question: given
$35^2 = 1225$, what is 36^2? Now, it would be altogether too
much labor to compute all the squares up to 35^2 even by
our clever system and to compute 34^2 in order to apply our
system would be as much labor as finding 36^2 by straight
multiplication. We need to pursue our investigation further. We had to add 5 to 2^2 in order to get 3^2, 15 to 7^2 in
order to get 8^2, 31 to 15^2 in order to get 16^2. Notice that
$5 = 2 \cdot 2 + 1$, $15 = 2 \cdot 7 + 1$, $31 = 2 \cdot 15 + 1$. The system thus seems to be: add one more than twice an integer
to get the square of the next higher integer. We now have
what seems to be a system for getting any square from its
predecessor. Will it always hold and, if so, why? No amount
of carrying out special cases in the form which we just considered would give us any clew as to the reason for its working in general. But we can begin to see what the general
situation is if we make use of the relationship between a
number and its successor as follows: $(15 + 1)^2 = (15 + 1)$
$(15 + 1) = 15(15 + 1) + 1(15 + 1) = 15^2 + 15 + 15 + 1$
$= 15^2 + 2 \cdot 15 + 1$. This is useful for it leads us to begin
to see that *whatever* the number used instead of 15 in $(15+1)^2$
the result would be that obtained by replacing 15 by that
number in $15^2 + 2 \cdot 15 + 1$. The easiest way to see this is
to let a stand for any number. Then $(a + 1)^2 = (a + 1)$
$(a+1) = a(a+1) + 1(a+1) = a^2 + a + a + 1 = a^2 + 2 \cdot a + 1$.

Using the letter a (or any other letter) has the advantage
that it points out that whatever number we had we could
deal with it exactly as we did with the a. It is a kind of
formula for the process of working it out. No matter what
numbers were put in for a at the beginning, the result would
be what one would get by replacing a by that number in
the conclusion. This, then, is an **algebraic proof** that to get
the square of 1 more than a certain number, one adds the

square of the certain number to 1 more than twice the number. Moreover, the algebraic proof tells us more than we started out to find; for any rational number, any real number, even any complex number, would give the same result, that is,

$$(a + 1)^2 = a^2 + 2a + 1.$$

Though geometry and algebra seem for the most part very distinct in high-school mathematics, they go hand in hand many times in proving certain results. It is enlightening to see that the above result can also be proved geometrically, that is, with the help of a picture. Consider the following arrays of dots:

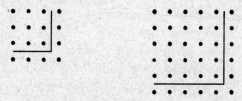

In the first array a little inspection shows that in order to get the 4 by 4 array from the 3 by 3 array we adjoin a column of three dots on the right, a row of three dots below and a dot in the lower right-hand corner. In other words, we add $2 \cdot 3 + 1$ dots. A similar situation prevails in the second array and can be seen to prevail in any square array of dots, by the following argument: suppose we have an n by n array of dots; if we adjoin at the bottom another row of n dots, at the right side another column of n dots, and in the lower right-hand corner one dot, we will have an $n + 1$ by $n + 1$ array of dots which has $n^2 + 2n + 1$ dots, whether n is 3 or 4 or whatever number of dots. And now we pause to ask whether this was a geometrical proof because we used a picture or an algebraic one because we used a letter n for any number in describing how the process would go in general. Of course, the answer is that it is both. But it certainly is not purely algebraic for we talked about dots. The proof is without doubt primarily geometric (the function of the algebra is merely to make our talk a little simpler) and we shall therefore call it a **geometric proof** even though we have used a little algebra.

Notice that by using the dots we have proved our result only for positive integers, for we could not speak of half a dot or -2 dots. There is another geometri-cal proof which will somewhat enlarge the set of numbers for which our result is true. Consider the square in Fig. 4:1. The area of the large square will be the area of the small one plus two rectangles a by 1 plus a square 1 by 1, which gives us the desired results.

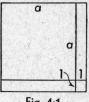

Fig. 4:1

EXERCISES

1. For what kinds of numbers a does the last proof apply?

2. What must be added to 3^2 to get 6^2, to 4^2 to get 7^2, in general to a^2 to get $(a + 3)^2$? Prove your result algebraically and geometrically, stating carefully for what kinds of numbers your results apply.

3. What must be added to a^2 to get $(a + b)^2$? Give a geometric and algebraic proof of your results.

4. Given the two numbers 8 and 3. Computation 1 is the following: $8 + 3 = 11$, $8 - 3 = 5$, $11 \cdot 5 = 55$. Computation 2 is the following: $8^2 = 64$, $3^2 = 9$, $64 - 9 = 55$. The two computations yield the same result. Prove algebraically and geometrically that this is true for any two positive integers. Will it hold for other kinds of numbers?

* 5. Prove that the sum of the digits of any perfect square (that is, a square of an integer) has one of the following remainders when divided by 9: 0, 1, 4, 7. Can you say anything about the order in which such remainders occur as one goes from one square to the next?

* 6. Consider Exercise 5 for cubes instead of squares.

* 7. Suppose we consider the cubes and, being wise, we write the differ-ences of successive cubes with them.

1	8	27	64	125	216	343	512	729	1000
	7	19	37	61	91	127	169	217	271

These differences do not behave as nicely as did the differences for the squares. How can one get from one cube to the next? Prove the formula that you get. Is there a geometric proof which you can find? Notice that, if you take the differences of the successive numbers of the second row of the table above, you get

12 18 24 30 36 42 48 54

which do behave well. Why do they behave as they do?

3. Triangular numbers and arithmetic progressions.

Suppose that instead of arranging our dots in squares we arranged them in tenpin fashion like this:

and so on. The numbers of dots in the successive triangles, assuming that a lone dot is triangular in shape, are:

$$1 \quad 3 \quad 6 \quad 10 \quad 15 \quad \cdots.$$

In a fashion analogous to our calling numbers in square array "square numbers" we call these **triangular numbers**. How can we get one triangular number from the preceding? The answer to this can easily be found, for we add 2 to the first triangular number to get the second, 3 to the second to get the third, $\cdots$, n to the $(n-1)$st to get the nth. This is, however, not really much help in finding the twentieth triangular number without finding the sum of

$$1 + 2 + 3 + 4 + 5 + \cdots + 17 + 18 + 19 + 20$$

which involves some labor. Though it is possible to find an algebraic derivation of a formula for the sum of the first n positive integers, it is easier first to find it geometrically. Furthermore, we have here to do with integers alone and there is no greater generality which an algebraic proof could give. Hence, we give a geometrical proof. Two sets of dots arranged in tenpin fashion can be made into the accompanying picture:

in which one array is turned upside down. Observe that in the total array there are six rows of dots and five slanting columns. Thus, there are $6 \cdot 5 = 30$ dots in all and hence $30/2 = 15$ dots in the triangular array. Similarly, if we represented by dots the twentieth triangular number we should have a triangle with 20 dots on a side. Forming two such triangles, inverting one and placing them together to form an array like that above, we should have 21 rows and 20 slanting columns, hence, 420 dots in all, which shows that in the triangular array there would be 210 dots. Therefore 210 is the twentieth triangular number. In general, if we represent the nth triangular number by a triangular array of dots and adjoin another such array inverted to form a figure like a parallelogram, we should have $n + 1$ rows and n slanting columns. Thus, *the nth triangular number is $n(n + 1)/2$.* This is equivalent to having the formula

$$(1) \quad 1 + 2 + 3 + \cdots + (n - 1) + n = n(n + 1)/2.$$

It is useful to notice what the above proof amounts to algebraically. If we write the sum above and place below it the same sum with the order reversed, we have

$$(2) \quad \begin{array}{c} 1 \; + \; 2 \; + \; 3 \; + \cdots + (n-2)+(n-1)+ \; n \\ n \; +(n-1)+(n-2)+ \cdots + \; 3 \; + \; 2 \; + \; 1 \\ \hline (1+n)+(1+n)+(1+n)+\cdots+(1+n)+(1+n)+(1+n)=n(1+n) \end{array}$$

Each $(1 + n)$ is the number of dots in one column of the double triangular array. As in the geometric proof, the sum of the numbers in the first line of (2) is thus $n(n + 1)/2$.

This algebraic trick (and, in fact, the geometric one) can be applied to give a more general result. Suppose we have any sequence of numbers beginning with the number a and such that the difference between two successive numbers is a fixed number d. Then they will be

$$a, a + d, a + 2d, \cdots, a + (n - 1)d = l$$

where there are n numbers in the sequence and l stands for the last one. This sequence is called an **arithmetic progression.** If we write the sequence in the reverse order and add as in (2) we have $a + l$ instead of $1 + n$ as the sum of each column and there are n columns, that is,

$$a \; +(a+d)+(a+2\,d)+\cdots+(l-2\,d)\;+(l-d)+\quad l$$
$$\underline{l \; +(l-d)\;+(l-2\,d)\;+\cdots+(a+2\,d)+(a+d)+\quad a}$$
$$(a+l)+(a+l)+(a+l)\quad+\cdots+(a+l)\quad+(a+l)+(a+l)$$

Hence,

$$(3)\quad a + (a + d) + (a + 2\,d) + \cdots + a + (n - 1)\,d = \frac{(a + l)n}{2}$$

where $l = a + (n - 1)\,d$, n is the number of terms in the sum and l is the last one. This reduces to formula (1) when $a = d = 1$.

In this connection, we quote a certain story about a famous mathematician, Gauss, which appears in E. T. Bell's *Men of Mathematics:*[1]

Shortly after his seventh birthday Gauss entered his first school, a squalid relic of the Middle Ages run by a virile brute, one Büttner, whose idea of teaching the hundred or so boys in his charge was to thrash them into such a state of terrified stupidity that they forgot their own names. More of the good old days for which sentimental reactionaries long. It was in this hell-hole that Gauss found his fortune.

Nothing extraordinary happened during the first two years. Then, in his tenth year, Gauss was admitted to the class in arithmetic. As it was the beginning class none of the boys had ever heard of an arithmetical progression. It was easy then for the heroic Büttner to give out a long problem in addition whose answer he could find by a formula in a few seconds. The problem was of the following sort, $81297 + 81495 + 81693 + \cdots + 100899$, where the step from one number to the next is the same all along (here 198), and a given number of terms (here 100) are to be added.

It was the custom of the school for the boy who first got the answer to lay his slate on the table: the next laid his slate on top of the first, and so on. Büttner had barely finished stating the problem when Gauss flung his slate on the table: "There it lies," he said — "Ligget se" in his peasant dialect. Then, for the ensuing hour, while the other boys toiled, he sat with his hands folded, favored now and then by a sarcastic glance from Büttner, who

[1] Published by Simon and Schuster, copyright 1937, by E. T. Bell.

imagined the youngest pupil in the class was just another block-head. At the end of the period Büttner looked over the slates. On Gauss' slate there appeared but a single number. To the end of his days Gauss loved to tell how the one number he had written was the correct answer and how all the others were wrong. Gauss had not been shown the trick for doing such problems rapidly. It is very ordinary once it is known, but for a boy of ten to find it instantaneously by himself is not so ordinary.

This opened the door through which Gauss passed on to immortality. Büttner was so astonished at what the boy of ten had done without instruction that he promptly redeemed himself and to at least one of his pupils became a humane teacher. Out of his own pocket he paid for the best textbook on arithmetic obtainable and presented it to Gauss. The boy flashed through the book. "He is beyond me," Büttner said; "I can teach him nothing more."

EXERCISES

1. Show that formula (1) above holds for $n = 8$.
2. Find the sums of the following arithmetic progressions:
   - a. $3 + 5 + 7 + \cdots$ for 20 terms.   Ans. 440.
   - b. $4 + \cdots + 39$ (having 13 terms).   Ans. $279\frac{1}{2}$.
   - c. $5 + 2 + \cdots - 19$.   Ans. -63.
   - d. $5 + 3 + \cdots$ for 19 terms.
   - e. $9 + 13 + \cdots + 41$.
   - f. $5 + \cdots + 53$ (having 15 terms).
   - g. $\frac{1}{2} - \frac{1}{3} - \cdots$ for 10 terms.
3. When 70 boards are piled vertically in order of length, each one is three inches longer than the one above it. The top one is one *foot* long. How long is the bottom one? If they were placed end to end in a straight line, what would be the combined length?
4. Are the figures given in the story about Gauss correct?
5. Show that the sum of the twelfth triangular number and the eleventh is equal to $12^2 = 144$.
6. Prove geometrically and algebraically that the sum of the $(a - 1)$st triangular number and the ath is equal to a^2.
7. Prove that the sum of the first n odd numbers is n^2 no matter what positive integer n is. What connection does this have with section 2?
8. Show that the differences of the squares of successive integers form an arithmetic progression. Can the same be said about the differences of the cubes of successive integers?
9. Give a geometric proof of formula (3) for $a = 2$, $d = 3$, $n = 5$.
10. What is the system for winning the following game: B mentions a

positive integer not greater than 5, A adds to it a positive integer not greater than 5, and so on? He who reaches 37 first, wins. If both B and A know the system and play it, who will win?

11. What would be the system for winning the above game if 5 and 37 were replaced by any other numbers?

12. Give the rule for working the following trick and explain your result: You choose a number and multiply it by 3. Tell me if the result is odd or even. If it is even, divide it by 2; if it is odd, add 1 to make it even and then divide by 2. Multiply your result by 3. Tell me what is the largest multiple of 9 less than your result. I will then tell you what number you originally chose.

* 13. Show that [1]
$$1^2 + 2^2 + 3^2 + \cdots + n^2 = \tfrac{1}{6} n(2n + 1)(n + 1).$$
* 14. Show that
$$1^3 + 2^3 + \cdots + n^3 = [n(n + 1)/2]^2.$$

4. Compound interest.

Though the results and formulas which we have derived have some uses outside of mathematics, we dealt with them largely as mathematical curiosities. There are, however, certain formulas which very obviously affect our daily lives. One who knows no algebra must take such formulas on faith — he must rely on the authority of those who know algebra.

In ancient times the taking of interest, "usury," was considered the lowest of practices but nowadays we all expect to receive something for the use of our money. In order that the reward shall be proportional to the amount loaned we charge a **rate of interest.** If the interest rate is 5% per year the interest on $100 is $5 and at the end of one year our $100 becomes $105; $1000 would accumulate interest of $50 in a year, and so on. In general, at an interest rate of i, P dollars will accumulate interest of Pi and thus, at the end of the year, P dollars becomes

$$A = P + Pi = P(1 + i).$$

This shows that, whatever the principal at the beginning of the year, the amount at the end of the year may be obtained

---

[1] See topic 3 at the close of the chapter.

by multiplying the principal by $(1 + i)$. If, now, the interest is *compounded annually* the total amount, $P(1 + i)$, at the end of the first year draws interest the second year (just as if it were drawn out and deposited again) and, by the above reasoning, we multiply this by $(1 + i)$ to get the amount at the end of the second year, that is,

$$P(1 + i)(1 + i) = P(1 + i)^2.$$

Similarly, if interest were compounded annually for three years, at the end of that time it would amount to $P(1 + i)^3$. Thus, we have shown that if interest is at the rate of i compounded annually, at the end of n years P dollars will amount to

(4) $$A = P(1 + i)^n.$$

This we call the **compound interest formula.** We call A the **amount** of P dollars at the interest rate of i compounded annually for n years. Thus, for example, \$100 at a rate of 3% compounded annually for 9 years would amount to

$$A = 100(1.03)^9 = 130.48.$$

One useful point of view is expressed by saying that \$100 now will be worth \$130.48 nine years from now if interest is 3% compounded annually.

Sometimes it is the amount rather than the principal that is known; as, for example, if I want to pay now a certain sum that at the end of nine years will amount to \$100. Then the expression is

$$100 = P(1.03)^9.$$

To find P, divide both sides of the equation by $(1.03)^9$ and have

$$P = \frac{100}{(1.03)^9} = \frac{100}{1.3048} = 76.64.$$

The compound interest formula may also be used when interest is compounded more often than once a year. For instance, if the interest rate is 6% compounded semiannually, the interest rate for each half year is 3%. Thus the amount of P dollars at 3% compounded annually for ten years is the same as the amount of P dollars at 6% compounded semiannually for five years since the rate per interest period is the same in both cases and the number of interest periods is the same. **Hence, formula (4) may be taken to be the amount of P dollars at the rate of i per interest period, for n interest periods.** For example, $100 compounded quarterly at 8% for ten years would be

$$A = P(1.02)^{40} = 220.80.$$

For the convenience of those computing interest, tables have been computed. We here give a small one which will be sufficient for our purposes.

$$A = (1 + i)^n$$

n	1%	2%	3%	4%	5%	6%	7%	8%
1	1.010000	1.020000	1.030000	1.040000	1.050000	1.060000	1.070000	1.080000
2	1.020100	1.040400	1.060900	1.081600	1.102500	1.123600	1.144900	1.166400
3	1.030301	1.061208	1.092727	1.124864	1.157625	1.191016	1.225043	1.259712
4	1.040604	1.082432	1.125509	1.169859	1.215506	1.262477	1.310796	1.360489
5	1.051010	1.104081	1.159274	1.216653	1.276282	1.338226	1.402552	1.469328
6	1.061520	1.126162	1.194052	1.265319	1.340096	1.418519	1.500730	1.586874
7	1.072135	1.148686	1.229874	1.315932	1.407100	1.503630	1.605781	1.713824
8	1.082857	1.171659	1.266770	1.368569	1.477455	1.593848	1.718186	1.850930
9	1.093685	1.195093	1.304773	1.423312	1.551328	1.689479	1.838459	1.999005
10	1.104622	1.218994	1.343916	1.480244	1.628895	1.790848	1.967151	2.158925

EXERCISES

1. I deposit $100 in a bank which pays interest of 2% compounded annually. What should I draw out at the end of ten years? *Ans.* $121.90.

2. I lend a friend $50. He promises to pay it back at the end of three years with interest at 6% compounded semiannually. How much should I receive?

3. What amount deposited in the bank of Exercise 1 will amount to $100 at the end of ten years? *Ans.* $82.03.

4. What should I receive now in total payment for a promise to pay $100 three years from now if interest is to be computed at 6% compounded semiannually?

5. Annuities and geometric progressions.

There is another problem connected with interest whose solution involves a little more mathematics. If you join a

Christmas Club you deposit a set amount of money at regular intervals throughout the year and just before Christmas receive from the bank an amount which, due to the interest, is more than the sum of the amounts which you deposited. This procedure is related to the answer for the following question: For how much money paid to you at present will you agree to pay to me or my heirs $100 a year for ten years? Such a series of payments constitute an **annuity certain**; an "annuity" because payments are annual (the term has come to include cases where payments are at any regular stated intervals), and "certain" because the number of payments does not depend on how long I live. Problems in other kinds of annuities are more complex and we shall not consider them at this point.

To simplify the discussion consider the following modification of the Christmas Club plan: P dollars is deposited at the beginning of each year for n years at an interest rate of i compounded annually. What will this amount to at the end of n years? To answer this question, notice that the first deposit will bear interest for n years, the second for $n - 1$ years, $\cdots$, the last for 1 year. Hence the amount will be

$$(5) \quad P(1 + i)^n + P(1 + i)^{n-1} + \cdots + P(1 + i)^2 + P(1 + i).$$

It is possible to simplify this sum. Notice that reading from the right, each term is obtained from the preceding by multiplying by $(1 + i)$. In the arithmetic progression we obtained each term by *adding* a fixed number to the preceding. A sequence of numbers such that each is obtained from the preceding by *multiplying* by a fixed number is called a **geometric progression**. (The adjective "geometric" occurs because the *ratio* of any number of the progression to the next one is the same as the ratio of the next to the one following that — ratio being an important geometrical concept.) In this case we thus need to find the sum of a sequence of numbers in geometric progression. Since the finding of such a sum is useful in many situations, we shall temporarily postpone our consideration of the problem of the Christmas Club and proceed to find a formula for the sum of any geometric progression.

To approach the problem gradually, let us first consider the progression

$$S = 3 + 6 + 12 + 24 + 48 + 96 + 192$$

which has the initial term 3 and the constant multiplier 2. It is not too laborious to add such a sum but it is not very enlightening, either. A trick is much more fun and instructive. Since 2 is the multiplier we multiply S by 2 and write it and the original sum carefully in the form

$$2S = \quad\quad 6 + 12 + 24 + 48 + 96 + 192 + 384$$
$$S = 3 + 6 + 12 + 24 + 48 + 96 + 192$$

The arrangement is an invitation to subtract and we have

$$2S - S = S = -3 + 384 = 381.$$

While this trick will give us the sum of any geometric progression, it is useful on occasion to have a formula for it. Thus, we now find the sum

$$S = a + ar + ar^2 + ar^3 + \cdots + ar^{n-1}.$$

Notice that this sum has n terms. Using the above trick we have

$$rS = \quad\quad ar + ar^2 + ar^3 + \cdots + ar^{n-1} + ar^n$$
$$S = a + ar + ar^2 + ar^3 + \cdots + ar^{n-1}$$

Subtraction gives

$$rS - S = -a + ar^n,$$
$$S(r - 1) = a(r^n - 1).$$

Divide both sides by $(r - 1)$ and have the formula for the sum of a geometric progression whose first term is a, whose ratio is r, and which has n terms:

$$(6) \quad\quad\quad S = a\left(\frac{r^n - 1}{r - 1}\right).$$

Another formula for the sum which is sometimes more convenient may be obtained by noticing that the last term in the

progression is $ar^{n-1} = l$. Then $ar^n - a = r(ar^{n-1}) - a = rl - a$.
And the formula may be written

(7) $$S = \frac{rl - a}{r - 1}.$$

We apply these formulas in the following examples.

Example 1. Find the sum of the geometric progression:

$$2 + 10 + 50 + 250 + \cdots + 6250.$$

SOLUTION: Since we know the last term but not the number of terms, we use formula (7) for the sum and have

$$S = \frac{5 \cdot 6250 - 2}{5 - 1} = \frac{31248}{4} = 7812.$$

Example 2. Find the sum of the first 15 terms of the geometric progression below. What is the last term?

$$2 + 6 + 18 + \cdots.$$

SOLUTION: Since we here know the number of terms we use formula (6) and have

$$S = 2 \cdot \frac{3^{15} - 1}{3 - 1} = 3^{15} - 1 = 14{,}348{,}906.$$

The last term is $2 \cdot 3^{14} = 9{,}565{,}938$.

Example 3. What should I receive four years from now in return for four annual payments of $100 beginning now if the interest rate is 3% compounded annually?

SOLUTION: The first payment bears interest for four years and hence, using the compound interest formula with $P = \$100$, $i = .03$, and $n = 4$, we see that at the end of the four years I should get $100(1.03)^4$ in return. Similarly for the second payment I should get $100(1.03)^3$ since it bears interest for three years. Continuing in this manner we have

$$A = 100(1.03)^4 + 100(1.03)^3 + 100(1.03)^2 + 100(1.03)$$

as the amount which I should receive at the end of the four years. This is a geometric progression, for each term may be obtained from the preceding by multiplying by 1/1.03. There is some gain in simplicity if we consider the order reversed and take the term 100(1.03) as the first term. Then the multiplier is 1.03, formula (6) for the sum of a geometric progression applies, and putting $a = 100(1.03)$, $r = 1.03$, and $n = 4$ we have

$$A = 100(1.03) \cdot \frac{(1.03)^4 - 1}{1.03 - 1} = 103 \frac{(1.03)^4 - 1}{.03}.$$

From the interest table we see that $(1.03)^4 = 1.125509$. Hence,

$$A = \frac{103(.125509)}{.03} = 430.91$$

where the answer is given to the nearest cent.

Example 4. Knowing that his daughter should be entering college five years from now, a man wishes to have on hand at that time $1000 to pay her expenses. To this end he proposes to make five equal annual payments beginning now. If he can get interest at 5% compounded annually, what should each payment be?

SOLUTION: Here we know the final amount but not the annual payment. Call the annual payment P. We see that the first payment would be worth $P(1.05)^5$ at the end of the five years, the second payment $P(1.05)^4$, and so on. They must in all amount to $1000 which gives us

$$1000 = P(1.05)^5 + P(1.05)^4 + P(1.05)^3 + P(1.05)^2 + P(1.05).$$

Our geometric progression now has $P(1.05)$ as the first term and $r = 1.05$. Thus,

$$1000 = P(1.05) \frac{(1.05)^5 - 1}{1.05 - 1} = P(1.05) \frac{1.276282 - 1}{.05}$$
$$1000 = P(5.801922)$$
$$\$172.36 = P.$$

Example 5. I wish to pay off a present loan of $1000 in five annual installments beginning a year from now. The rate of interest is 5%. There are two plans by which it can be done. Plan A: At the end of each year I pay one-fifth of the principal (that is, $200) plus the interest on that portion of the principal unpaid at the beginning of that year (that is, the **outstanding principal**). Plan B: The annual payments are to be equal. Find the annual payments under each plan.

SOLUTION: For Plan A we can best show the solution by making a table:

	Principal unpaid at beginning of year	Payment on principal	Interest paid	Principal unpaid at end of year
First year	1000	200	50	800
Second year	800	200	40	600
Third year	600	200	30	400
Fourth year	400	200	20	200
Fifth year	200	200	10	0

Notice that the interest paid over the five years is $150. The payments under this plan are easy to compute but it has the disadvantage that the first year I have to pay $250 in all, which is $40 more than the last payment.

Plan B: Why is it that $230 is not the annual payment under this plan? You should be able to answer that question before carrying through the computation. Let P be the annual payment. Since the first payment is made a year from now, its value now is $P/(1.05)$. The second payment is worth now $P/(1.05)^2$, and so on. Their total present values must be $1000. Thus we have

$$1000 = \frac{P}{1.05} + \frac{P}{(1.05)^2} + \frac{P}{(1.05)^3} + \frac{P}{(1.05)^4} + \frac{P}{(1.05)^5}.$$

This is a geometric progression in which the first term is $P/(1.05)^5$ and the multiplier is 1.05. Hence,

$$1000 = \frac{P}{(1.05)^5} \frac{(1.05)^5 - 1}{1.05 - 1} = \frac{P(1.276282 - 1)}{1.05^5(.05)},$$
$$1000 = P(4.32948),$$
$$230.97 = P.$$

Notice that the total amount of interest paid in this case is almost $5 more than in plan A. It is instructive to form a table for this plan by way of comparison. Plan B is often referred to as "amortization of a debt." The word means literally "killing off."

Amortization Table for a Debt of $1000 Paid in Five Years at 5 %

	First year	Second year	Third year	Fourth year	Fifth year
Principal unpaid beginning of year	1000.00	819.03	629.01	429.49	219.99
Annual payment	230.97	230.97	230.97	230.97	230.97
Interest	50.00	40.95	31.45	21.47	11.00
Payment on principal	180.97	190.02	199.52	209.50	219.97
Principal outstanding at end of year	819.03	629.01	429.49	219.99	.02

The computation proceeds in this fashion: in the first column, the interest is 5% of the $1000. The difference between this and the annual payment ($180.97 in the first column) goes toward payment of the principal. This latter amount subtracted from the top entry in the column gives the principal unpaid at the end of the year. Notice that 5% of the principal outstanding at the end of any year is the interest entry for the next year. The fact that two cents is still outstanding at the end of the five years results from the necessity to make each entry in the table to the nearest cent. In this table the entries for a year are in one column to make the computation easier. If the payments were made over a longer period of years it would be necessary to have a wider sheet of paper or change the arrangement.

EXERCISES

1. Find the value of the following sums:

   a. $6 + 18 + 54 + \cdots + 486.$ *Ans.* 726.

   b. $3 + 12 + 48 + \cdots + 12288.$

2. In one second a certain living cell divides to form two cells, in the next second each of these two divides to form two more, and so on. Estimate the number of cells at the end of a minute. If at the end of a minute the cells exactly fill a thimble, how full will the thimble have been at the end of 59 seconds? Assume all cells are of the same size.

3. There is an old story which, with several variations, appears in

D. E. Smith's *History of Mathematics*.[1] A certain Arab king was so pleased with the invention of the game of chess that he told its inventor that he would grant any request. Whereupon the inventor asked for one grain of wheat for the first square, two for the second, four for the third, eight for the fourth, and so on in geometric progression until the 64th square was reached. Assuming that there are fifteen million grains of wheat to the ton, show that the inventor asked for more than one hundred billion tons (the number "billion" being used in the United States sense). Compare the number of grains for the 20th square with the sum of the numbers of grains for the first nineteen squares.

4. In the case of each of the following, state whether it is an arithmetic progression, a geometric progression, or neither. Find the sum of the first ten terms if it is either of the two named progressions:

a. $\frac{1}{3} + \frac{1}{2} + \frac{2}{3} + \frac{5}{6} \cdots$.

b. $\frac{1}{3} + \frac{1}{6} + \frac{1}{12} + \frac{1}{24} + \cdots$.

c. $\frac{1}{3} + \frac{1}{4} + \frac{1}{5} + \frac{1}{6} + \cdots$.

d. $\frac{1}{2} - \frac{1}{3} + \frac{2}{9} - \frac{4}{27} + \cdots$.

e. $\frac{2}{5} + \frac{1}{15} - \frac{4}{15} - \frac{9}{15} - \cdots$.

5. A gambler plays a number of games. If he wins the first he gets 1 cent, if he loses he pays 1 cent. The gain or loss in the second game is 2 cents, in the third 4 cents, and so forth, the gain or loss each time being double that for the previous time. If his total capital is 2^{20} cents (about $10,000) and he loses every time, how many games can he play and still pay his losses? If he loses the first thirteen games and wins the fourteenth what will be his net gain or loss?

6. The sum of $10 is deposited at the beginning of each year for five years. If interest is paid at the rate of 4% per year compounded annually what should be the amount at the end of five years? What should be the amount at the beginning of the fifth year immediately after the fifth deposit has been made?   *Ans.* $56.33, $54.16.

7. Find the annual payment and construct a table similar to that for Example 5 for the amortization of a debt of $1000 at 4% over a period of six years.   *Partial ans.* The annual payment, $190.76.

8. I give to a bank a government bond which will mature to $1000 five years from now (that is, its value five years from now will be $1000). In return I expect the bank to pay me five equal yearly payments beginning at present. If interest is 3% compounded annually, how much should each of the payments be?   *Ans.* $182.87.

9. In return for what sum deposited now should a bank give ten semi-annual payments of $100 beginning six months from now? Interest is 4% compounded semiannually.   *Ans.* $898.26.

10. A man purchases a farm for $15,000 agreeing to pay $5000 cash

[1] Reference **32**, pp. 549 ff.

and the balance (principal and interest) in ten equal annual payments beginning a year from the date of purchase. What should each payment be if interest is 5% compounded annually? Compare your result with that of Example 5, Plan B.

11. Mr. Rounds wishes to accumulate a fund which will amount to $1000 at the end of ten years. In order to do this he makes ten equal annual payments. Find the amount of each payment, first if it is made at the beginning of each year, second if at the end of each year. Interest is 4% compounded annually.

12. What amount deposited now in a bank paying 2% interest compounded semiannually would entitle the depositor to ten semiannual withdrawals of $1000 each beginning six months from now?

13. Mr. Wood bought an automobile, paying $1000 cash and $1000 at the end of each year for three years. What would be the cash price of the car if money is worth 6% compounded annually?

14. On January 1, 1922, a streetcar company issued $10,000 in 5% bonds to mature January 1, 1932. (That is, the bonds pay 5% of the face value each year and the total face value at the time of maturity.) If the bonds are to be redeemed by sinking funds at 4% interest, how much must be set aside from the company's earnings at the end of each year to provide for the interest and the retirement of the debt?

15. There are two plans under which a man can pay off a $10,000 mortgage in ten equal annual payments with interest at 5% compounded annually. Plan A: he pays off principal and interest in five equal annual installments. Plan B: at the end of each year he pays the interest for that year on $10,000, namely $200. At the same time he deposits an amount in a savings bank giving 5% interest, the amount being so determined that at the end of ten years he will have $10,000 in the savings bank to pay off the principal. Is the total annual payment under the two plans the same? If so, or if not, why?

16. In return for a loan of $1000 at the beginning of a certain year, a man agrees to pay at the end of each year $200 plus the interest at the rate of 6% on the balance remaining due at the beginning of the year. What is the total amount which he pays? Does this problem have to do with an arithmetic or geometric progression or neither?

17. Show that the value of the sum: $1 + 1/2 + \cdots + 1/2^n$ is

$$2 - (1/2)^n.$$

18. For the geometric progression 6, 18, 54, 162, $\cdots$ the differences of successive numbers, 12, 36, 108, $\cdots$ also form a geometric progression. Show that this will be true for every geometric progression. Will the progression of the differences ever be the same as the original progression? If so, when?

19. Is there a result for arithmetic progressions analogous to that in Exercise 18?

20. If *a b c* are three numbers in arithmetic progression, what is *b* in terms of *a* and *c*?

21. If *a b c* are three numbers in geometric progression, what is *b* in terms of *a* and *c*?

Note: The *b* in Exercise 20 is called the **arithmetic mean** (or average) of *a* and *c* while in Exercise 21 it is called the **geometric mean** of *a* and *c*, or mean proportional of *a* and *c*.

* 22. Prove that the arithmetic mean of two positive numbers is never less than the geometric mean of those numbers.

* 23. Given a rectangle. Prove that the side of a square of equal perimeter is never less than the side of a square of equal area.

* 24. Porky, a guinea pig, has a litter of 7 female pigs on each birthday. Each of her offspring does likewise. Assuming that none die, how many pigs will start life on the day when Porky is three years old? How many descendants will she have immediately after she is five years old? Formulas for the answers may be given and the answers estimated.

6. The binomial theorem.

There is a close connection between calculation of such an interest table as we have been using and a certain theorem which turns out to be useful in numerous situations. The so-called **binomial theorem** is in reality a formula for $(x + y)^n$. (The expression $x + y$ is called a **binomial** because it has *two* terms in it.) At this stage we are not concerned with getting a formula but merely in developing a quick method of calculating the result for small values of n.

To this end, let us see what $(x + y)^n$ looks like for $n = 2$, 3, and 4.

$$(x + y)^2 = x(x + y) + y(x + y) = \begin{aligned} x^2 + \quad & xy \\ + \quad & xy + y^2 \\ \hline x^2 + 2\,xy & + y^2 \end{aligned}$$

$$(x + y)^3 = x(x^2 + 2\,xy + y^2) + y(x^2 + 2\,xy + y^2)$$
$$= \begin{aligned} x^3 + 2\,x^2y + \quad & xy^2 \\ + \quad x^2y + 2\,xy^2 & + y^3 \\ \hline x^3 + 3\,x^2y + 3\,xy^2 & + y^3 \end{aligned}$$

$$(x + y)^4 = \begin{aligned} x^4 + 3\,x^3y + 3\,x^2y^2 + \quad & xy^3 \\ + \quad x^3y + 3\,x^2y^2 + 3\,xy^3 & + y^4 \\ \hline x^4 + 4\,x^3y + 6\,x^2y^2 + 4\,xy^3 & + y^4 \end{aligned}$$

We can shorten this process if we use merely the coefficients of the various expressions, that is, the numbers which multiply the various terms. For $(x + y)$ we write 1 1 and get the coefficients of $(x + y)^2$ by adding as follows:

$$\begin{array}{ccc} 1 & 1 & \\ & 1 & 1 \\ \hline 1 & 2 & 1 \end{array}$$

Similarly, to get the coefficients of $(x + y)^3$ we write

$$\begin{array}{cccc} 1 & 2 & 1 & \\ & 1 & 2 & 1 \\ \hline 1 & 3 & 3 & 1 \end{array}$$

Furthermore, if we knew the coefficients of $(x + y)^7$ to be

$$1 \quad 7 \quad 21 \quad 35 \quad 35 \quad 21 \quad 7 \quad 1$$

we could get the coefficients of $(x + y)^8$ as follows:

$$\begin{array}{ccccccccc} 1 & 7 & 21 & 35 & 35 & 21 & 7 & 1 & \\ & 1 & 7 & 21 & 35 & 35 & 21 & 7 & 1 \\ \hline 1 & 8 & 28 & 56 & 70 & 56 & 28 & 8 & 1 \end{array}$$

We can avoid writing the coefficients of $(x + y)^7$ a second time by using the following scheme:

$$\begin{array}{ccccccccc} 1 & 7 & 21 & 35 & 35 & 21 & 7 & 1 \\ 1 & 8 & 28 & 56 & 70 & 56 & 28 & 8 & 1 \end{array}$$

where now each number in the second line except the first and last is the sum of the two nearest numbers in the line above it. By this method we can compute the coefficients of $(x + y)^n$ for successive values of n.

If, then, to complete the picture we call $(x+y)^0 = 1$ we have the following table of coefficients for $n = 0, 1, 2, \cdots, 9, 10$.

n											
0						1					
1					1		1				
2				1		2		1			
3			1		3		3		1		
4		1		4		6		4		1	
5	1		5		10		10		5		1
6	1		6		15		20		15		6		1
7	1		7		21		35		35		21		7		1
8	1		8		28		56		70		56		28		8		1
9	1		9		36		84		126		126		84		36		9		1
10	1		10		45		120		210		252		210		120		45		10		1

This picture goes by the name of the **Pascal triangle**, though it was known to a Chinese algebraist in 1300, over three centuries before Pascal's day. We shall in a later chapter find formulas for all the terms in the expression for $(x + y)^n$.

Now let us apply our newly found knowledge to the computation of an entry in the interest table. The coefficients of the expression for $(x + y)^9$ are found in the next to the last line of the Pascal triangle, and putting $x = 1$ and $y = .03$ we have

$$(1.03)^9 = 1 + 9(.03) + 36(.03)^2 + 84(.03)^3 \\ + 126(.03)^4 + 126(.03)^5 + \cdots.$$

Notice that $(.03)^4 = .00000081$ and $126(.03)^4 = .00010206$. The sixth term is $.0000030618$. Since after the sixth term the coefficients (84, 36, 9, 1) become smaller and smaller we can see that each term after the sixth is less than $.03$ times the previous term. Hence the sum of the first five terms will give us a value of $(1.03)^9$ which is accurate to *five* decimal places and the sum of the first six terms gives accuracy to six places. If we are to find the amount for a principal of $1000, five-place accuracy (that is, five terms) will give us accuracy to cents in the value of the amount; and the first six terms would suffice for a principal of $10,000. Taking the sum, then, of the first six terms we have the result that $(1.03)^9 = 1.304773$ accurate to six decimal places.

It should be pointed out that the fact that the binomial theorem is of some assistance in this connection depends vitally on the fact that y in $(x + y)^9$ is small. Fortunately (at least for purposes of this calculation) interest rates are small. If y were 1 the successive terms would, in the beginning, increase.

Considered on its own merits, the Pascal triangle has many interesting properties. For purposes of description let us speak of the slanting lines parallel to the sides as being the *diagonals* of the triangle and number them from the outside inward. Thus 1 1 1 1 $\cdots$ is the first diagonal, 1 2 3 4 $\cdots$ is the second, and so forth. Notice that the numbers of the third diagonal are apparently the triangular numbers (see Exercise 5 below). The numbers in the fourth diagonal are what are called **pyramidal numbers,** for if cannon balls, for instance, are piled in a triangular pyramid the smallest such pyramid will contain one ball, the next will contain 4, the next 10, etc. (see Exercise 6 below).

EXERCISES

1. Use the binomial theorem to find the amount of $100 at 2% interest compounded annually for ten years; for six years.

2. Find the coefficients of the powers of x and y in the expansion of $(x + y)^{12}$. Use this result to answer the following: Find the amount of $100 at the end of one year if interest is 6% compounded monthly.

Ans. $106.17.

(Notice that this is little larger than would be the amount if interest were compounded annually.)

3. Prove that the differences of successive numbers in any diagonal of the Pascal triangle are the numbers after 1 in the previous diagonal. (For instance, in the third diagonal $3 - 1 = 2$, $6 - 3 = 3$, $10 - 6 = 4$, $\cdots$ and $2, 3, 4, \cdots$ are the numbers after 1 in the second diagonal.)

4. A United States Savings Bond purchased January 1, 1940, for $75 has from that date until maturity the values given in the following table:

	1940	1941	1942	1943	1944	1945	1946	1947	1948	1949	1950
Jan. 1	$75.00	75.50	76.50	78.00	80.00	82.00	84.00	88.00	92.00	96.00	100.00
July 1	$75.00	76.00	77.00	79.00	81.00	83.00	86.00	90.00	94.00	98.00	

Show that the interest rate earned between purchase and maturity is a little less than 3% compounded semiannually. During what half-year is the rate of interest greatest? Compute the rates of interest for each half-year and show that their average is 2.9% per year. Should this be the average rate for the ten years?

* 5. Prove that the numbers in the third diagonal of the Pascal triangle are triangular numbers and use your result to find a formula for the third term in the expression for $(x + y)^n$.

Hint: Notice, for instance, that $10 = 4 + 3 + 2 + 1$ and show that each number in the third diagonal is the sum of the numbers in the previous rows of the second diagonal.

* 6. Prove that the numbers in the fourth diagonal of the Pascal triangle are pyramidal numbers.

* 7. Prove that the formula for the fourth term in the expression for $(x + y)^n$ is $n(n - 1)(n - 2)/6$.

** 8. Prove that the sum of the numbers in the nth row of the Pascal triangle is equal to 2^n.

7. Nonterminating progressions.

There is another application of geometric progressions which can well be illustrated by an ancient paradox which is very well known: the story of Achilles and the tortoise. The argument, you recall, was that Achilles could never catch the tortoise because before reaching the tortoise he would have to traverse half the distance between him and the tortoise and by that time the tortoise would have moved farther along; the process continues without end "showing" that Achilles could never catch the tortoise. As a matter of fact, the same argument would apply if the tortoise remained stationary. For simplicity's sake we, therefore, assume the tortoise was asleep and that he did not walk in his sleep. Further, suppose that when the story begins Achilles is 4 miles from the tortoise. He must traverse half the distance, that is, 2 miles, then half the remaining distance, namely, 1 mile and, continuing, we have a sum like this:

$$2 + 1 + 1/2 + 1/4 + \cdots$$

each term being obtained from the preceding by taking half of it. Notice that after having gone 2 miles he had 2 left to go, after having gone $2 + 1$ he had 1 left to go, after $2 + 1 + 1/2$ he had $1/2$ left to go, and so on, after having gone $2 + 1 + 1/2 + \cdots + 1/2^n$ he had $1/2^n$ left to go. The farther one goes in the sequence the less there is left to go. More precisely, $2 + 1 + 1/2 + \cdots + 1/2^n$ differs from 4 by $1/2^n$. In fact, the progress of Achilles may be seen from the accompanying diagram:

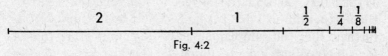

Fig. 4:2

From this we see that, by taking n large enough, our sum will differ from 4 by as little as we please; not only that, but for all values of n larger than "large enough" it will differ from 4 by the same or less than "as little as we please."

For example, if you select the number $1/1000$, I take $n = 10$ and not only will the sum then differ from 4 by less than $1/1000$ but for all values of n greater than 10 the sum will differ from 4 by less than $1/1000$; if you select the number $1/1,000,000$, I take $n = 20$ and not only will the sum then differ from 4 by less than $1/1,000,000$ for this value of n but also for all greater values of n. In fact, no matter how small a number you name, I can take n so large that for that value of n and for all greater values of n the sum will differ from 4 by less than that amount. We express this by saying that **the limit** *of* $2 + 1 + \cdots + 1/2^n$ *as n increases without bound* is 4. And we write it, for short,

$$\lim_{n \to \infty} 2 + 1 + 1/2 + \cdots + 1/2^n = 4.$$

We could express the same result in the language used for repeating decimals by saying

$$2 + 1 + 1/2 + 1/4 + \cdots$$

converges to 4. Notice that 5 is not the limit because though it is true that as n increases, the sum comes closer and closer to 5, yet, if you gave me the number $1/2$, I could not choose n so large that the sum would differ from 5 by less than $1/2$. As a matter of fact, the sum differs from 5 by more than 1 for all values of n. On the other hand, $3\frac{7}{8}$ is not the limit, for though it is true that if you select, for instance, the number $1/1,000,000$, I can choose n so large, namely $n = 3$, that the sum will differ from $3\frac{7}{8}$ by less than $1/1,000,000$; but it will not be true that for all values of n greater than 3 the sum will differ from $3\frac{7}{8}$ by less than $1/1,000,000$. As a matter of fact for all values of n greater than 3 the sum will differ from $3\frac{7}{8}$ by at least $1/16$. Notice, further, that the sum cannot have any limit other than 4, for suppose that it had a limit N. Then if you name the number $1/1,000,000$ I must be able to find a number n greater than 20 such that the sum differs from N by less than $1/1,000,000$. We know from the above that the sum differs from 4 by less than $1/1,000,000$. Hence N differs from 4 by less than 2/1,000,000. We can use this process to show that N differs from 4 by as small an amount as you please and the only number N which has this property is 4 itself.

One can use the formula for the sum of a geometric progression to see that the limit of the sum is 4. By Exercise 17 at the close of section 5 we have $1 + 1/2 + 1/4 + \cdots + 1/2^n = 2 - (1/2)^n$ and hence $2 + 1 + 1/2 + \cdots + 1/2^n = 4 - (1/2)^n$, which shows what we previously have noticed, namely, that the sum differs from 4 by $1/2^n$.

The misconception in the "paradox" of Achilles and the tortoise is, of course, that since the distance which Achilles travels in catching the tortoise is indeed $2 + 1 + 1/2 + \cdots$ without end it is supposed, without justification, that the time required is also without end. We can actually compute the time. Suppose that Achilles was walking at the rate of four miles an hour and that, as before, the tortoise was fast asleep and four miles away. It would take Achilles 1/2 hour to walk the 2 miles, 1/4 of an hour to walk the 1 mile, 1/8 to walk the 1/2 mile, and so forth. Thus the time which he would take to walk $2 + 1 + 1/2 + 1/4 + \cdots + 1/2^n$ miles would be $1/2 + 1/4 + \cdots + (1/4)(1/2^n)$ of an hour. As n becomes larger and larger the time comes closer and closer to 1. As a matter of fact

$$\lim_{n \to \infty} [1/2 + 1/4 + \cdots + (1/4)(1/2^n)] = 1.$$

This shows that, as we knew all the time, the time taken to catch the tortoise is one hour.

EXERCISES

1. Show that $\lim_{n \to \infty} 1 + 1/3 + 1/9 + \cdots + 1/3^n = 3/2$. For what values of n will the sum differ from 3/2 by less than 1/1,000,000?

2. Show that $\lim_{n \to \infty} 1 + x + x^2 + \cdots + x^n = \dfrac{1}{1 - x}$ for all real numbers x such that $-1 < x < 1$.

3. Find the $\lim_{n \to \infty} (1/10 + 1/100 + \cdots + 1/10^n)$. What does this show about the unending decimal .1111111 $\cdots$?

4. What fraction has the decimal value .12121212 ⋯?

SOLUTION: The result of Exercise 2 is here helpful though we could go back to the formula for the sum of a geometric progression. Write

$$.121212 \cdots = 12(.01 + .01^2 + .01^3 + \cdots)$$
$$= 12(.01)(1 + .01 + .01^2 + \cdots)$$
$$= .12 \lim_{n \to \infty} (1 + .01 + .01^2 + \cdots + .01^n).$$

Putting $x = .01$ in the result of Exercise 2 we have

$$.121212 \cdots = .12 \frac{1}{1 - .01} = \frac{12}{100 - 1} = \frac{12}{99} = \frac{4}{33}.$$

A second method of solution is the following which uses the same trick we used in finding the formula for the sum of a geometric progression. Let $x = .121212 \cdots$ (we know from our statement in Chapter III that there is such a number). Then

$$100\,x = 12.121212 \cdots$$
$$x = .121212 \cdots$$

and subtraction gives us

$$99\,x = 12$$

and hence

$$x = 12/99 = 4/33.$$

5. What fraction has the decimal value .123123123 ⋯?

6. What fraction has the decimal value 56.101101101 ⋯?

7. Show that $\lim_{n \to \infty} [1/2 - 1/4 + 1/8 - 1/16 + \cdots - (-1/2)^n] = 1/3$.

8. Does the sum $1 - 1 + 1 - 1 + \cdots - (-1)^n$ have a limit as n increases without bound?

9. Does the sum $1 + 2 + 4 + 8 + \cdots + 2^n$ have a limit as n increases without bound?

10. Prove $\lim_{n \to \infty} (n - 1)/n = 1$.

11. A ball is dropped from a height of 16 feet. After first striking the ground it rises to the height of $(3/4) \cdot 16$ feet, on the second bounce it rises to three-fourths the height it rose on the first bounce, and so it continues to rise on each bounce three-fourths the height it rose on the previous one. Assume that it bounces vertically and find how far the ball travels before coming to rest. How long will it take? (Assume the distance traveled by a falling body in time t to be $s = 16\,t^2$.)

12. A boy in a swing moves from north to south in an arc of 12 feet, then from south to north in an arc of 9 feet, and so on, each time swinging through an arc three-fourths the length of the previous swing. How far will he travel? If he swings strictly in accordance with the above statement and if each swing takes just as long as the next, will he ever come to rest? Give your reasons.

* 13. Prove that every repeating unending decimal has a rational number as a limit, that is, is the expansion of a rational number.

* 14. Let S_n be a sum of n numbers such that $\lim\limits_{n \to \infty} S_n = A$ and such that S_n is, for each value of n, greater than S_n for the preceding value of n. Show that if for $n = B$, S_n differs from A by less than a number c, then S_n differs from A by less than that number c for *all* values of n greater than B.

8. The ways of equations.

Before we proceed to further applications of algebra we shall do well to delve into the ways of equations. An algebraic equation is an equality between two algebraic expressions. Some equations involve only numbers (e.g., $5/3 = 10/6$) while some involve letters as well as numbers. In the latter case it is sometimes true that no matter what numbers we substitute for the letters, the equation still holds. Examples of this type of equation are

$$3\,x = 5\,x - 2\,x,$$
$$x^- - y^2 = (x - y)(x + y).$$

These are called **identical equations** or **identities,** and the left side is said to be *identically* equal to the right side. Other equations do not hold for all numerical values which one can give to the letters. Such equations are called **conditional equations.** Examples of this type of equation are

$$3\,x = 3\,x^2/x$$
$$x + 2 = 3$$
$$x + y = 3$$
$$0 \cdot x = 1.$$

The first of these examples holds except when $x = 0$, the second only for $x = 1$, the third only for those values of x and y whose sum is 3, and the last for no values of x.

In the equations which we deal with here, the letters stand for numbers and we work with them as we do with numbers. The commutative law then enables us to say $xy = yx$ identically, the distributive law to say $x(y + z) = xy + xz$ identically, and so forth. Since our letters stand for numbers, we freely replace any expression by any other identically equal to it.

By a **solution of an equation** we mean a set of values of the letters involved for which the equation holds. For example, $x = 3$, $y = 0$ is a solution of $x + y = 3$. We have seen that an equation may have many solutions, a few, or none at all.

The usual method of solving an equation is to derive from it a series of other equations until finally we have one whose solution is evident. Of course, there should be some connection between the solution of the given equation and that which forms the final step. In many problems we can so manage it that each equation of the series is **equivalent** to that which precedes it, that is, has the same solutions — all and no others. For instance, $x/2 = 3$ and $x = 6$ are equivalent equations since every solution of one is a solution of the other. This is so because we multiply the first equation by 2 to get the second, showing that every solution of the first is a solution of the second; also we can divide the second equation by 2 to get the first, showing that every solution of the second is a solution of the first. However, to multiply $x/2 = 3$ by 0 yields the equation $0 \cdot x = 0$ which is not equivalent, since one cannot get from the second to the first by dividing by 0; furthermore, there are many solutions of the second equation which are not solutions of the first, though every solution of the first is one of the second.

The following processes yield equivalent equations:

1. *Multiplying or dividing both sides of an equation by some number different from 0.*

2. *Subtracting or adding the same number to both sides of an equation.*

The above statement is back of our solving the equation $3x + 7 = 3$ first by subtracting 7 from both sides, getting $3x = -4$ and second by dividing both sides by 3, getting $x = -4/3$. Since each equation is equivalent to its predecessor, the last is equivalent to the first. Therefore $x = -4/3$ is the solution and the only solution of the first equation.

The two equations $a = b$ and $a^2 = b^2$ are not equivalent, for, though every value of a and b which satisfies the former equation satisfies the latter, it is not true that every solution of the latter is a solution of the former. The equation $a^2 = b^2$ merely implies that $a = b$ *or* $a = -b$. Thus, squaring both sides of an equation *does not* yield an equivalent equation. This has a bearing on the usual method of solution of an equation like $\sqrt{x + 1} = 3$. We square both sides and get $1 + x = 9$ or $x = 8$. The latter equation is not equivalent to the given one and we must substitute the final value into the given equation to see whether or not it is a solution. In this case $x = 8$ is indeed a solution but if the original equation had been $\sqrt{x + 1} = -3$, then $x = 8$ would not be a solution since the radical sign means the positive square root.

The following principle is then important: *unless, in the process of solving an equation, each equation is equivalent to its predecessor, one cannot be sure that all solutions of the last equation are solutions of the first.*

Of course, we usually manage it so that in the series of equations every solution of an equation is a solution of its successor, and hence that all the solutions of the given equation are among those of the final equation. In that case we need try in the given equation only the solutions of the final equation. In the example above, we knew that the solutions of $x = 8$ include the solutions of $\sqrt{x + 1} = 3$ and also of $\sqrt{x + 1} = -3$.

In dealing with equations, you should beware of artificial rules of thumb which go by the names of "canceling" or "transposing." For example, we may have $x + 2 = 4$ and conclude from it that $x = 4 - 2 = 2$. This process, of course, amounts to "transposing 2 and changing the sign" but it is just as brief, much more sensible, and much safer to say "subtract 2 from both sides." One reason that the latter is safer than the former is that we are apt to forget what "transpose" means but we are not apt to forget what "subtract" means. For instance, in $x/2 = 4$ we can "transpose" the 2 and change the sign to get either $x = 4/-2$ or $x = 4 - 2$, neither of which is correct. By the time we put all the necessary restrictions on the word "transpose" we might just as well have said "subtract" and be done with it. Similarly $3 x = 12$ implies $x = 4$ because we can *divide* both sides by 3.

EXERCISES

Find all the solutions of each of the following equations. State exactly what you are doing in each step of the process and point out what lead to equivalent equations and what do not.

1. $5 x + 6 = 3$.

2. $\dfrac{x}{3} - 8 = 20$.

3. $\dfrac{3}{x - 2} = 5$.

4. $x + 2 = 2 x + 2 - x$.

5. $\sqrt{x - 5} = 3$.

6. $\sqrt{x - 5} = -3$.

7. $\dfrac{3}{x - 2} = \dfrac{10}{2 x - 4}$.

8. $1000 = 2.357 P$.

9. $100 = \dfrac{2.462 P}{.03}$.

9. The antifreeze formula.

This situation often arises in the winter months to those of us who own cars: I have enough alcohol in the radiator of my car for a temperature of 10°, and the radiator is full. The newspaper says that tonight the temperature is going to drop to 0°, the newspaper is always right, and I have no garage for my car. How much alcohol must I put in after drawing an equal amount of mixed water and alcohol from my radiator? Usually, lacking sufficient data and energy, we rely on our own guess or that of the garage man who might "play safe," that is, guess too high. But it is possible to find the exact answer without much difficulty. Of course, we are helpless unless sometime when the garage man is not looking we copy his table of the amount of alcohol required for various temperatures. It would look something like the first two columns of the following table, where the first column gives the freezing temperature when a tank of 15-quart capacity contains the amount of "alcohol" (really about 90% alcohol) listed in the second column, the amount given a little more accurately than is likely to be given in his table. We have also listed the percentage in the third column.

Temperature, degrees Fahrenheit	Quarts of "alcohol" in 15-quart tank	Percentage of "alcohol"
−20°	7.7	51
−10°	6.6	44
0°	5.7	38
10°	4.2	28
20°	3.0	20

Quarts are given to the nearest tenth and the percentages to the nearest unit.

Now the guessing begins. You might say, "Since I have 4.2 quarts and need 5.7 quarts, I should add 1.5 quarts." The garage man says, "Better put in two quarts." And you are both wrong. For suppose you draw off 1.5 quarts of the mixture, you then have .28(13.5) quarts of alcohol in your radiator, that is, 3.8 quarts. If you add 1.5 quarts to that, you have 5.3 quarts which is short of the 5.7 required. On the other hand, if you follow the advice of the garage man and draw off 2 quarts you have .28(13) quarts of alcohol in your radiator, that is, 3.6 quarts. Adding 2 quarts to that gives 5.6 which is (we fooled you) again less than the 5.7 quarts required. This shows that we misjudged the garage man. You may well ask, "What is a quart of alcohol among friends?" But why avoid algebra! With a good formula we can get an accurate answer in less time than it took to go through the above computations.

Postponing the derivation of the formula, we can solve the problem at hand by letting x be the amount in quarts of the 28% mixture which must be withdrawn and replaced by alcohol. What is left after the withdrawal will be .28(15 − x) quarts of alcohol. After adding x quarts of alcohol we will have x + .28(15 − x) quarts of alcohol in our 15 quarts. Since we want a 38% mixture we require .38(15) quarts of alcohol in the resulting mixture, and hence we have

$$x + .28(15 - x) = .38(15)$$
$$x + .28(15) - .28\,x = .38(15)$$
$$x - .28\,x = .38(15) - .28(15)$$
$$(1 - .28)x = (.38 - .28)15$$
$$.72\,x = .1(15) = 1.5$$
$$x = 2.1$$

This tells us that 2.1 quarts are actually needed.

It should be noted that in combining two substances in a solution there is always a certain amount of shrinkage. It turns out to be true that there is little shrinkage with alcohol and water. At the other extreme, it is well known that a cupful of sugar put into a cupful of water yields just about one cupful of sweet water.

As it was pointed out, the alcohol is not pure and the table is made taking this into account; the percentages are percentages of 90% alcohol. In the exercises below, unless something is said to the contrary, the alcohol is of the strength provided for in the table.

EXERCISES

1. A full 15-quart radiator has just enough alcohol for $-10°$. How much should be withdrawn and replaced by alcohol to have just enough alcohol for $-20°$? What would your answer be to the question if the radiator held 30 quarts?

2. A radiator whose capacity is V quarts is full of $s\%$ solution of alcohol. How much of the solution should one withdraw and replace with alcohol to have a full radiator of $S\%$ alcohol?

3. How could computation by the preceding formula be simplified by making use of the usual chart giving the number of quarts of alcohol required for various temperatures and various radiator capacities? Use this to compute the answers for Exercise 1.

(Notice that the results of the last two exercises with the table of solutions required for various temperatures gives us a quick means of solving all such antifreeze problems. Henceforth, to get our answer for any particular case we need merely substitute the proper values for the letters in the formula.)

4. How much of 10 gallons of a 10% solution of alcohol in water should be withdrawn and replaced by a 40% solution to yield 10 gallons of a 20% solution? How much of a 40% solution should be *added* (no withdrawal) to 10 gallons of a 10% solution to yield a 20% solution?

5. A certain company finding that it has a surplus of $2000 decides to distribute it among its workers in the form of a bonus. Seeking to give greater reward for long service it divides its workers into three classes as follows: Class 1: those who have been employees for less than five years; Class 2: those not in Class 1 who have been employees for less than ten years; Class 3: all other employees. The bonus of each of those in the second class is to be twice that of each of the' first and of each of those in the third class three times that of each in the first. If there are 20 employees in the first class, 50 in the second, and 40 in the third, how should the bonus be divided?

6. In the situation of Exercise 5, the surplus is P dollars; the numbers in the three classes are n_1, n_2, and n_3, respectively. Find a formula for the amounts received by the workers.

7. In a certain state, the income tax is 4% of the *net income*, that is, the income after all deductions are made. One of the allowable deductions is the amount of the tax itself. If I is the income after all deductions are made except that of the tax itself, what, in terms of I, is the amount of the tax?

10. Puzzle problems.

There is no hard and fast line between solving equations and finding formulas, but there is some difference in point of view. In the last section, in deriving each formula we solved innumerable equations once and for all by finding a formula for the result. This has a great advantage if we are going to solve a large number of problems of the same kind, but none whatever in solving one. It is chiefly in the puzzle type of problem that we are interested in a result for a single numerical problem. We do not need a general formula because if we have solved one puzzle problem we have no further interest in any other problem which differs from it only in the numbers involved. In such problems we are told something about a certain number — given various conditions about it — and wish to find what number or numbers satisfy these conditions. Hence we let some letter be the number and so manipulate the conditions that we find what number the letter must be. As a matter of fact, we did this in the last section when we found the antifreeze formula.

For purposes of illustration consider the following problem which is so simple that it can be solved without any use of algebra, but which we shall solve in an algebraic way to bring the method into relief: what three consecutive integers have 27 as their sum? To answer this we suppose that there are three consecutive integers whose sum is 27. The second would be 1 more than the first, the third 2 more than the first, and the sum of the three would be 3 more than 3 times the first. Since 27 is 3 more than thrice the first, $24 = 27 - 3$ must be thrice the first, and the first number is $24/3 = 8$. In that case the others would be 9 and 10. Now $8 + 9 + 10 = 27$ and we have indeed found three consecutive integers whose sum is 27. Furthermore, they are the only three consecutive integers whose sum is 27 since we showed above that the first must be 8 if there are any.

This whole process can be expressed more concisely if, after assuming that there are three such integers, we let y be the first one. Then the others are $y + 1$ and $y + 2$ and we have $y + y + 1 + y + 2 = 3y + 3 = 27$. Thus $3y = 24$ and, dividing both sides by 3, we have $y = 8$. Each equation is equivalent to its predecessor and hence $y = 8$ is the solution and the only solution of the original equation. Thus the three consecutive numbers are 8, 9, 10.

However, suppose we wish the answer to the following question: What number beside zero is its own double? Let x be such a number and $x \neq 0$. Then $x = 2x$ and we might proceed to its solution as follows: $x^2 = 4x^2$ and hence $0 = 4x^2 - x^2 = 2x - x$. That is, $3x^2 = x$. We are assuming that x is not zero and hence can divide both sides by x having $3x = 1$ and $x = 1/3$. We have thus shown that if there is any number different from zero which is its own double, it is 1/3, for in our sequence of equations the solution of each equation is included among the solutions of its successor. But since our last equation is not equivalent to the first one it may have, and in fact does have, a solution which does not satisfy the original equation. The last equation does show, however, that since $x = 1/3$ does not satisfy the given equation, there is no value of x different from zero which will. It is enlightening to attempt to retrace the sequence of equations. Now $3x = 1$ implies $3x^2 = x$ which in turn implies that $4x^2 - x^2 = 2x - x$ but this does not imply that $2x = x$.

EXERCISES

1. Is there any solution to the following? If so, find it. Find four consecutive integers whose sum is 28. State very carefully the steps in the argument.

2. In attempting to solve the following problem the method given below might be used which leads to a value which does not satisfy the original condition. At what point does the retracing of the argument first break down?

Problem: Find a number whose positive square root is 1 more than the positive square root of 1 more than the number. Let x be the number and have

$$\sqrt{x} = \sqrt{x+1} + 1.$$

Square both sides to get

$$x = x + 1 + 2\sqrt{x+1} + 1$$
$$0 = 2 + 2\sqrt{x+1}$$
$$-2 = 2\sqrt{x+1}$$
$$-1 = \sqrt{x+1}.$$

Square both sides again and have

$$1 = x + 1. \qquad \text{Thus } x = 0.$$

3. To answer the question, what numbers are equal to their cubes, we set up the equation $x^3 = x$. Dividing by x we get $x^2 = 1$ and $x = 1$ or -1. Both the numbers 1 and -1 are equal to their cubes but 0 is also equal to its own cube and does not appear in the final result. Why?

4. Is there an integer such that if it is subtracted from its cube the result is 1 less than the integer? If so, find it.

5. Christopher is now twice as old as Phyllis. Two years ago he was three times as old as Phyllis. How old are they now?

SOLUTION 1: Let p be Phyllis' age. Then $2p$ is Christopher's age. Two years ago Phyllis was $p - 2$ years old and Christopher was $2p - 2$. Since Christopher was then three times as old as Phyllis, we have

$$2p - 2 = 3(p - 2),$$
$$2p - 2 = 3p - 6.$$

Add 6 to both sides and subtract $2\,p$ from both sides to get $4 = p$. Each equation is equivalent to its predecessor, which shows that Phyllis is 4 years old and Christopher 8.

SOLUTION 2: Let p be Phyllis' age and c Christopher's age. The first sentence of the statement of the problem expressed algebraically is then

$$c = 2\,p.$$

The children's ages two years ago were $c - 2$ and $p - 2$. Then the second sentence of the problem may be written

$$c - 2 = 3(p - 2).$$

Add 2 to both sides and get

$$c = 2 + 3(p - 2) = 2 + 3\,p - 6 = 3\,p - 4.$$

This, taken with $c = 2\,p$, gives

$$2\,p = 3\,p - 4.$$

Add 4 to both sides and subtract $2\,p$ to get

$$4 = p.$$

Logically, we have shown that if there are any ages satisfying the conditions imposed, we have found them. Though it is possible here to consider equivalent equations, the simplest way to show that we have a solution is to try the numbers in the problem as given. Christopher, now being 8, is twice as old as Phyllis. Two years ago they were 6 and 2 respectively, which satisfies the other condition.

6. A bottle and a cork together cost $1.05. The bottle costs exactly $1.00 more than the cork. How much does the cork cost?

7.[1] Weary Willie went up a certain hill at the rate of $1\frac{1}{2}$ miles per hour and came down at the rate of $4\frac{1}{2}$ miles per hour, so that it took him just 6 hours to make the double journey. Now how far was it to the top of the hill?

8.[2] A man bought two cars but, due to an unforseen circumstance, found that he had to dispose of them. He sold them for $1200 each, taking a loss of 20% on one and making a profit of 20% on the other. Did he make a profit on the whole transaction or a loss and to what extent?

9.[3] In a recent motor ride it was found that we had gone at the rate of 10 miles an hour, but we made the return journey over the same route at 15 miles an hour, because the roads were more clear of traffic. What was our average speed?

10.[4] A man persuaded Weary Willie, with some difficulty, to try to work on a job for 30 days at 8 shillings a day, on the condition that he

---

[1] Puzzle 28, reference **14**.
[2] Adapted from puzzle 9 in reference **12**.
[3] Puzzle 67 in reference **12**.
[4] Puzzle 10 of reference **14**.

would forfeit 10 shillings a day for every day that he idled (that is, the net loss for each idle day was to be 2 shillings). At the end of the month neither owed the other anything, which entirely convinced Willie of the folly of labor. Now, can you tell just how many days' work he put in and on how many days he idled?   *Ans.* He worked 6 days.

11.[1] "When I got to the station this morning," said Harold Tompkins, at his club, "I found I was short of cash. I spent just one-half of what I had on my railway ticket, and then bought a penny newspaper. When I got to the terminus I spent half of what I had left and twopence more on a telegram. Then I spent half of the remainder on a bus and gave threepence to that old matchseller outside the club. Consequently, I arrive here with this single penny. Now, how much did I start with?"
                                                       Ans. 42 pence.

12.[2] A meeting of the Amalgamated Society of Itinerant Askers ⋯ was held to decide whether the members should strike for reduced hours and larger donations. It was arranged that during the count those in favor of the motion should remain standing, and those who voted against should sit down.

"Gentlemen," said the chairman in due course, "I have the pleasure to announce that the motion is carried by a majority equal to exactly a quarter of the opposition." (Loud cheers.)

"Excuse me, guv'nor," shouted a man at the back, "but some of us over here couldn't sit down."

"Why not?"

" 'Cause there ain't enough chairs."

"Then perhaps those who wanted to sit down but couldn't will hold up their hands ⋯. I find there are a dozen of you, so the motion is lost by a majority of one." (Hisses and disorder.)

Now how many members voted at that meeting?

SOLUTION: We can let N be the total number in the meeting and S the number standing. Then

(i) $$S = 5(N - S)/4.$$

The number really wanting to vote for the motion is $S - 12$ and those wanting to vote against would be $N - S + 12$. Thus

(ii) $$S - 12 + 1 = N - S + 12.$$

The first equation gives $4S = 5N - 5S$ or $9S = 5N$ and the second gives $2S - 23 = N$. Thus $9S = 5(2S - 23) = 10S - 115$. Hence $S = 115$ and $230 - 23 = 207 = N$. Thus we have shown that if there is any solution to the problem $N = 207$ and $S = 115$. Substitution shows that these values satisfy equations (i) and (ii).

[1] Puzzle 9 in reference **15**.
[2] Puzzle 83 in reference **14**.

13. The combined ages of Mary and Ann are 44 years, and Mary is twice as old as Ann was / when Mary was half as old as Ann will be / when Ann is three times as old as Mary was / when Mary was three times as old as Ann. How old is Mary? — Sam Lloyd.

SOLUTION: Divide the statement of the problem by the / as shown and let m and a be the ages of Mary and Ann respectively. Then we have

(i) $$m + a = 44.$$

Suppose before the first / Ann was $a - y$ years old. Then

(ii) $$m = 2(a - y).$$

At that time Mary was $m - y$ years old and let $a + z$ be the age Ann will be before the second /. We then have

(iii) $$m - y = (a + z)/2.$$

Let $m - t$ be Mary's age before the third / and have

(iv) $$a + z = 3(m - t).$$

Ann was then $a - t$ years old and we have

(v) $$m - t = 3(a - t).$$

At this point you are expected to bring out your paper and pencil and do what you are told. Solve the last equation for t and get $t = (3\,a - m)/2$. Substitute this in equation (iv) and have $a + z = 9(m - a)/2$. Substituting this in equation (iii) gives $m - y = 9(m - a)/4$. Solve this for y to get $y = (9\,a - 5\,m)/4$. Substitute this in equation (ii) and have $m = 2\,a - (9\,a - 5\,m)/2$ which implies $5\,a = 3\,m$ and $a = 3\,m/5$. Substitute this value of a in equation (i) and have $m + 3\,m/5 = 44$. Hence $8\,m = 220$, $m = 55/2 = 27\frac{1}{2}$ and $a = 16\frac{1}{2}$. Notice that there was a definite system to the process. We put in new variables as we needed them. Then, working from the end, we systematically eliminated them.

We have thus shown that if there is a solution the values are what we have found. Now the easiest way to show that there is a solution in this case is to show that the values we have found satisfy the conditions of the problem. To do this we work from the end of the problem toward the front. Equation (v) expresses the fact that t years ago Mary was three times as old as Ann. We found that $t = (3\,a - m)/2$ and $a = 33/2$ and $m = 55/2$ gives $t = 11$ and we see that $55/2 - 11 = 3(33/2 - 11)$. Thus, when Mary was three times as old as Ann she was $33/2$ years old. Ann's age after the second / was then $3 \cdot 33/2 = 99/2$ years. After the first /, then, Mary was $99/4$ years old. How old was Ann then? The answer to this question is that, since that was $55/2 - 99/4 = 11/4$ years ago, Ann must have been $33/2 - 11/4 = 55/4$ years old. And, sure enough, Mary is twice as old as that.

We could, of course, have demonstrated that we have indeed a solution by showing that every step in its derivation is reversible. But the process of the last paragraph not only proves that we have found a solution but checks the derivation itself.

14. A fish's head is 9 inches long. His tail is as big as his head plus one-half his body. His body is as big as his head and tail together. How long is the fish?

15. A ship is twice as old as the boiler was when the ship was as old as the boiler is. The total age of the ship and boiler is 49 years. How old is the ship?   *Ans.* 28.

16. Louisa is 2 years older than Christopher. Six years ago she was twice as old as he. What are their ages now?

17. Two years ago Louisa's age was the sum of the ages of Christopher and Phyllis. Two years from now Louisa will be twice as old as Phyllis and the ages of all three will total 28 years. What are their ages?

18. Mary is 24 years old. She is twice as old as Ann was when she was as old as Ann is now. How old is Ann?

19.[1] Ten years ago a man married a widow; they each already had children. Today there was a pitched battle engaging the present family of twelve children. The mother ran to the father and cried, "Come at once! Your children and my children are fighting our children!" As the parents now had each nine children of their own, how many were born during the ten years?

* 20. The following is a considerable simplification of a type of problem which the tax departments of certain corporations in Pennsylvania must solve every year. Let I be the given income and S be the amount of the combined capital stock, surplus, undivided profits, etc., before all income taxes are computed. Let X be the amount of the Pennsylvania income tax, Y the federal income tax, and Z the Pennsylvania capital stock tax. We have the following:

$$X = .07(I - X - Y - Z)$$
$$Y = .19(I - X - Z)$$
$$Z = .005(S - X - Y - Z).$$

Find a formula for X, Y, and Z in terms of I and S.

The following three problems differ from those above in that not enough data are given to determine all the unknowns in the statement, though enough is given to answer the question proposed. A clever person can discover short cuts which may, in some cases, eliminate the use of algebra altogether.

* 21. A man rowing upstream passed a bottle floating down with the current and noticed that, when he passed it, it was opposite a certain stump. After he had rowed steadily upstream for 45 minutes it occurred to him that the bottle might have something in it and he immediately

---

[1] Puzzle 47 in reference **15**.

turned around and rowed downstream after the bottle, catching up with it three miles below the stump. Assuming that the man rowed at a constant rate relative to the stream, that the stream flowed at a constant rate, and that the bottle did not catch on any snags, find how fast the stream was flowing.

* 22. Mr. X, a suburbanite, catches the 5 o'clock train for home daily and is met at his station by his chauffeur. One day he took the 4 o'clock train. On arriving at his station he decided to start walking for home without phoning for his chauffeur. His chauffeur started for the station at the customary time, met Mr. X on the way, and returned home with him. They arrived home 20 minutes before their usual time. How long did Mr. X walk?

* 23.[1] If an army 40 miles long advances 40 miles while a dispatch rider gallops from the rear to the front, delivers a dispatch to the commanding general, and returns to the rear, how far has he to travel?

Ans. $40(1 + \sqrt{2})$.

11. Diophantine equations.

So far in the problems we have been considering, it was sometimes true that an answer which was not a positive integer would have been senseless, but the problem was so set up that the result turned out to be a positive integer. There are certain types of puzzle problems in which there would be an unlimited number of answers were it not for the restriction 'that they must be positive integers. Such problems give rise to equations whose solutions are restricted to be integers; such equations are called **Diophantine equations,** from Diophantus, the Greek, who was one of the first to study them.

Consider the following such problem: A girl goes to a candy store and spends $1.00 on candy. Gumdrops are ten for a cent, caramels are 2¢ apiece, and chocolate bars are 5¢ each. She confines herself to these three kinds of candy and buys in all 100 pieces of candy. How many of each kind does she get? To answer this let g be the number of gumdrops, c the number of caramels, and b the number of chocolate bars. Since there are 100 pieces,

$$(8) \qquad\qquad g + c + b = 100.$$

---

[1] Puzzle 38 of reference **14.**

Since they cost $1.00 we have

(9) $$g/10 + 2\,c + 5\,b = 100.$$

Solving for g in equation (8) gives $g = 100 - c - b$ and substituting in equation (9) we have

$$\frac{100 - c - b}{10} + 2\,c + 5\,b = 100.$$

Multiply by 10 to get

(10) $$100 - c - b + 20\,c + 50\,b = 1000.$$

This gives

(11) $$19\,c + 49\,b = 900.$$

We shall consider two systematic methods for the solution of equations like (11). Both begin with the observation that if integers b and c can be found satisfying equation (11) then $19\,c = 900 - 49\,b$ which is 0 on a circle of 19 divisions; that is, we wish to determine an integer b so that $900 - 49\,b$ is 0 on this circle. Now, on such a circle 900 is 7 since $900 = 19 \cdot 47 + 7$ and 49 is 11. Hence, we wish to choose b so that $7 - 11\,b$ is 0 on this circle. At this point the two methods diverge.

1. *The brute force method* is to try $b = 1, 2, \cdots$ until you find some value which makes $7 - 11\,b$ have the value 0 on the circle of 19 divisions. At worst it would be necessary to try 18 different values of b. (Why is this so?) In this case $b = 11$ is the least positive value which works. Now $b = 11$ on the circle means that b can be 11, 30, 49, $\cdots$ or -8, -27, $\cdots$ off the circle. A short way of writing this is to set $b = 11 + 19\,k$, where k is an integer (positive, negative, or zero). However, b and c are positive numbers and equation (11) shows that $49\,b$ must be less than 900. This would not be the case if b were 30 or had any greater value and b is not negative. Hence $b = 11$ and $c = 19$ is the only possible solution in positive integers of equation (11). Then equation (8) shows that $g = 70$. These values satisfy equation (9) and our solution is complete.

2. *The trickery method* shortens the labor but requires a little ingenuity. As above, we have $11\,b = 7$ on the circle with 19 divisions. We seek to reduce the multiplier of b. The first step is then to write $-8\,b = 7$ on the circle. We can further reduce the multiplier if we replace 7 by -12 and have $-8\,b = -12$ or $8\,b = 12$. Divide both sides by 4 to get $2\,b = 3$, make the right side even again by replacing 3 by -16 and $2\,b = -16$ implies $b = -8$ or $b = 11$.

One remark should be made about the observation before the discussion of the two methods. We could, of course, have written $49\,b = 900 - 19\,c$ and talked about the circle of 49 divisions, but the numbers are larger and our solution would be longer.

A problem of this kind, to be "good," should be so drawn up that there is one solution and only one. A carelessly made problem is apt to have many or no solutions. For instance, a problem which led to the equation $5\,x + 3\,y = 12$ would have no solution in positive integers x and y while $5\,x + 3\,y = 38$ has three solutions. The reader should satisfy himself that these statements are true.

EXERCISES

1. The following Chinese problem of the sixth century is given by David Eugene Smith (reference **32**, p. 585). If a cock is worth 5 sapeks, a hen 3 sapeks, and 3 chickens together 1 sapek, how many cocks, hens, and chickens, 100 in all, will together be worth 100 sapeks? Find that solution in which the number of chickens is the largest.

Ans. 12 cocks, 4 hens, 84 chickens.

2. A farmer with $100 goes to market to buy 100 head of stock. Prices were as follows: calves, $10 each; pigs, $3 each; chickens, $0.50 each. He gets exactly 100 head for his $100. How many of each does he buy?

3. Three chickens and one duck sold for as much as two geese; one chicken, two ducks, and three geese were sold together for 25 shillings. What was the price of each bird in an exact number of shillings?[1]

Ans. Chickens, 2; ducks, 4; geese, 5 shillings.

4. A man received a check for a certain amount of money, but on cashing it the cashier mistook the number of dollars for the number of

[1] Puzzle 19 in reference **15**.

cents and conversely. Not noticing this, the man then spent 68 cents and discovered to his amazement that he had twice as much money as the check was originally drawn for. Determine the amount of money for which the check must have been written.[1]

5. Johnny has 15 cents which he decides to spend on candy. Jelly beans are seven for a cent, licorice strips three for 2 cents, and caramels a cent apiece. He gets 40 pieces of candy. How many of each kind does he buy?

6. Percival has 50 cents to spend on candy. In *his* candy store jelly beans are ten for 9 cents, licorice strips eleven for 10 cents, and caramels 2 cents apiece. He wants to get just as many pieces of candy as he can regardless of kind. How many of each does he purchase?

7. A man's age at death was 1/29 of the year of his birth. How old was he in the year 1900?[2]

8. Find all the solutions of

$$35\,x + 19\,y = 372$$

where x and y are positive integers.   *Ans.* $x = 9, y = 3$.

9. Find all positive integer values of x, y, and z which satisfy both the equations

$$x + 3\,y - 4\,z - 8 = 0,$$
$$2\,x + y + 3\,z - 39 = 0.$$

10. A man bought an odd lot of wine in barrels and one barrel containing beer. There were, in all, six barrels containing 15, 16, 18, 19, 20, 31 gallons respectively. He sold a quantity of the wine to one man and twice the quantity to another, but kept the beer to himself. None of the barrels was opened. Which barrel contained the beer?[3]

11. The equation 16/64 = 1/4 is a result which can be obtained by the cancellation of the 6 in numerator and denominator. Find all the cases in which $ab/bc = a/c$ for a, b, and c integers between 1 and 9 inclusive.

* 12. The number 3025 has the property that if the number 30 and 25, formed of the digits in order of the two halves of the number, are added (giving 55) and the result squared (55²) the result is the original number. How many other numbers are there composed of four different digits having the same property?[4]

** 13. Five women, each accompanied by her daughter, enter a shop to buy cloth. Each of them buys as many feet of cloth as she pays cents per foot. In each case the number of feet bought and the number of cents spent is a positive integer.

---

[1] **35**, p. 65.
[2] Puzzle 52 in reference **15**.
[3] Puzzle 76 in reference **12**.
[4] Rewording of puzzle 113 in reference **12**.

a. Each mother spends $4.05 more than her daughter.
b. Mrs. Evans buys 23 feet less than Mary.
c. Rose spends 9 times as much as Clara.
d. Mrs. Brown spends most of all.
e. Effie buys 8 feet less than 10 times as much as Mrs. White.
f. Mrs. Connor spends $40.32 more than Mrs. Smith. The other girl's name is Margaret. What is her surname?

* 14. In *Kitty Foyle*, by Christopher Morley, appears the following story:[1]

That clock was made in imitation of Horticultural Hall, you could see all the works, which were usually slow. "It has Philadelphia blood," Pop said. The kitchen clock, a much more homely thing, was fast. Once we had an argument, how long would it be, with the two clocks getting farther apart from each other at a definite rate, before they would both again tell the same time. Mac said that was easy and went upstairs to figure it out by algebra. Mother said they never would because two wrongs never make a right. Pop went in the back yard to think it over. Mac came downstairs once to ask, did we mean till both clocks told the same time or till they both told the correct time?

"The correct time," Pop said. "Gee," Mac said, "I bet that'll be about a thousand years." He went upstairs again to calculate, and didn't come back at all. Meanwhile they forgot to wind the clock and it stopped.

Assume that at noon both clocks told exactly the correct time and that the Horticultural Hall clock loses time at the constant rate of 3 minutes per hour and the kitchen clock gains at the rate of 4 minutes per hour. When will both clocks next tell exactly the correct time (provided they are regularly wound)? Could the rates of losing and gaining be such that both clocks would never again agree exactly — much less tell the correct time together?

12. Topics for further study.

1. History of algebra: See reference **32**, Chap. VI. In particular, the Pascal triangle is discussed on pp. 508–511.

2. Triangular, square and, more generally, pentagonal, hexagonal and other "polygonal" numbers: See references **35**, pp. 5–11; **23**, pp. 201–216.

3. Sums of series (cf. Exercises 13 and 14 of section 3): See references **28**, Chap. X; **35**, pp. 5–11.

4. Classification of various annuity problems and derivation of formulas for each.

[1] From Kitty Foyle, pp. 28–29. Copyright, 1939, by Christopher Morley. Published by J. B. Lippincott Co., Philadelphia.

5. Limits: See reference **29,** Chap. 9.

6. Various puzzles: See references **5, 12, 13, 14, 15, 26.**

7. The complete solution in integers of the Diophantine equation $a^2 + b^2 = c^2$ (a, b, and c are integers). Notice that a and b may be considered to be the legs of a right triangle and c the hypotenuse. See reference **11,** Chap. 1; also **28,** Chap. 4.

CHAPTER V

Graphs and Averages

I gotta love for Angela,
    I love Carlotta, too.
I no can marry both o' dem,
    So w'at I gona do?

Oh, Angela ese pretta girl,
She gotta hair so black, so curl,
An' teeth so white as anytheeng.
An' oh, she gotta voice to seeng,
Dat mak' your hearta feel eet must
Jomp up an' dance or eet weell bust.
An' alla time she seeng, her eyes
Dey smila like Italia's skies,
An' makin' flirtin' looks at you —
But dat ees all w'at she can do.

Carlotta ees no gotta song,
But she ees twice so big an' strong
As Angela, an' she no look
So beautiful — but she can cook.
You oughta see her carry wood!
I tal you w'at, eet do you good.
W'en she ees be som'body's wife
She worka hard, you bat my life!
She nevva gettin' tired, too —
But dat ees all w'at she can do.

Oh, my! I weesh dat Angela
    Was strong for carry wood,
Or else Carlotta gotta song
    An' looka pretta good.
I gotta love for Angela,
    I love Carlotta, too.
I no can marry both o' dem,
    So w'at I gonna do? [1]

---

[1] "Between Two Loves" from *Carmina* by T. A. Daly. Reprinted by permission of Harcourt, Brace and Company, Inc.

1. Measurements.

No doubt Carlotta had black hair, too, but it was not so black and it may have been curly but perhaps it did not curl much. She could probably even sing, but it might have been more enjoyable when she did not attempt it. On the other hand, Angela could surely carry wood but not much. This all goes to show that the answer to "how much?" is at least as important as the answer to "what?" is.

Different persons (as well as one person) are interested in different things. When water is the topic, the physicist is interested in its composition and the degree to which it will combine with other substances; the thirsty traveler, in its purity (if he is not too thirsty). Once the property in which one is interested is isolated, the important thing to determine, if one is to have a complete description, is to what degree this property is possessed. We know that the answer to "how many?" is given by stating a number. This answer is so satisfactory that we immediately want to answer the question of "how much?" or "to what degree?" by assigning a number to it. But this is much harder to do. The beauty of Angela's song or the savor of Carlotta's cooking could not be measured in terms of mere numbers.

Description needs adjectives and most adjectives imply a comparison. For instance, to say that a star is very bright merely implies that it is brighter than most stars which we can see. Such a statement is not as definite as "it is brighter than Sirius"; then, we begin to have a measure of its brightness. Since comparison plays a fundamental role in measurement and consists in listing things in order, we pause for a moment to consider the **properties of order** in terms of the brightness of a star. They are:

1. If A and B are any two stars, star A is just one of three things: it is either brighter than B, just as bright as B, or not so bright as B.

2. If A is brighter than B, then B is not so bright as A and vice versa. If A is just as bright as B, then B is just as bright as A.

3. If A is brighter than B, and B is brighter than C, then A is brighter than C. The same is true if "brighter than" is replaced by "not so bright as" or by "just as bright as."

Since brightness has these three properties we say that the stars **can be ordered according to their brightness.**

It is clear that when we have the properties of order we can arrange our objects according to the degree to which a certain quality appears. This is a step in measurement, but it has its disadvantages for one must have the whole list more or less in mind in order to give meaning to a certain position in that list. The next step is to fix upon a **unit of measurement.** This is done in two essentially different ways.

One way of fixing the unit is to decide arbitrarily on one thing which shall be said to have one unit of the quality. Just as soon as we have something which weighs one pound we can express all weights by numbers with reference to that unit of weight. Anything which weighs twice as much (that is, as much as two things which each weigh one pound) we say weighs two pounds, anything which weighs half as much weighs half a pound, and so forth. By comparing our weights with a single one we have the advantage of not only having arranged our objects in an order according to the given quality, but have a definite measure in terms of a number. A knowledge of the unit of weight would enable me, for instance, to tell my friend in Australia how heavy my piano is without taking it over for him to compare with his piano. It should be remarked in this connection that the accuracy of such a system of measurement depends on the care with which the unit is defined, and its usefulness depends on its availability. To the former end the International Bureau of Weights and Measures was founded in Paris in 1875 for the purpose of agreeing upon certain international units of measurement. The units of measurement in this country are defined with reference to certain units in the National Bureau of Standards. The standard pound, for example, is a cylinder of pure platinum about 1.35 inches high and 1.15 inches in diameter. There are also certain legal definitions of units.

For instance, the international conference agreed that the alternative and provisional definition of a metre should be 1,553,164.13 times the wave length of the red light emitted by a cadmium vapor lamp excited under certain specified conditions. (The legal definition is apt to stand up better under bombing than could the platinum cylinder.) The availability of units usually is promoted through the use of certain instruments of measurement graduated with reference to the standard unit or a secondary unit obtained from it.

A second way to fix upon a unit is with reference to two extremes. This is often necessary, for in many cases the first method of defining a unit fails. To say that Angela is twice as beautiful as Carlotta or that one pan of water is twice as hot as another has no meaning (at least until a unit is defined) because two Carlottas are no more beautiful than one and two pans of lukewarm water are not hotter than one. In the case of Angela and Carlotta we might decide that no one could be more beautiful than the former, and we will call her beauty 100 and that Carlotta had no beauty whatsoever and her beauty would be 0. Our unit of beauty would then be one-hundredth of the difference between, and we should at least have the basis for its numerical measurement. In the case of water, no water is hotter than steam and hence we call it 100; and none is colder than ice and hence we call it 0. This is the basis of the centigrade scale and is used, of course, not only for measuring the temperature of water but for all temperatures. To be accurate we must further specify that the basis of measurement shall be for pure (that is, distilled) water at sea level. Notice that "twice as hot" in the centigrade scale is very different from "twice as hot" in the Fahrenheit scale, but "twice as heavy" is the same no matter what units are used.

The first kind of unit described is called an **absolute unit** for it is determined with reference to an absolute zero, that is, a condition in which the property concerned is completely lacking. The pound is determined relative to absence of weight. Similarly if one measures temperature in absolute units, that is, with reference to absolute zero, then "twice as hot" does have a meaning and this is independent of the unit chosen. Once one could agree on what was absolute lack of beauty, one could then, independent of the unit chosen, give meaning to "twice as beautiful."

These units are quite arbitrary, as may be judged from the wide variety of units in this world. But a unit, to be usable, must be comparable in magnitude with the quality which it measures. Vegetables are measured in pounds but aspirin in grains.

After determining units of measure one devises instruments for measuring: scales for weight, the thermometer for heat, and goodness knows what for beauty. The more accurate the instrument, the more accurately is our comparison established.

EXERCISES

1. According to which of the following properties can objects be ordered: weight, size, temperature, color, distance, value, loudness, place of manufacture? If, in any cases, your answers are debatable, give a short debate. In the cases in which the answer is unequivocably "yes," by what must one replace "brighter than" in the three properties of order above?

2. Name at least one unit for each of the following qualities: length, area, strength of a person, specific gravity, amount of electric current, brightness, excellence of work in a course, hardness of jewels, strength of a salt solution. (It is expected that students will, if necessary, supplement their knowledge by the use of dictionaries and encyclopedias to answer these questions.) In which cases is the unit an absolute unit?

3. Describe briefly instruments used to measure at least three of the qualities listed above.

2. Graphs of sets of measurements.

It was pointed out in Chapter III that different numbers could, with advantage, be thought of as corresponding to points on a line. But graphical representation is chiefly useful in presenting the relationship between two sets of values. It depicts such relationships in the large more vividly than does a mere table of numbers.

We are all more or less familiar with the way in which a graph is drawn, but it may be worth while to consider one case in detail. Suppose our table of values is

Table 1

t	0	1	2	3
s	0	16	64	144

where t is the time in seconds and s is the distance in feet traversed by a falling body, the distance being measured from the position of the body when $t = 0$. We draw a pair of perpendicular lines and choose the horizontal line on which to represent the values of t and the vertical line those of s. We call the former line the **t-axis** and the latter the **s-axis,** and their point of intersection the **origin.** Then on each of the lines we choose the values of t and s at the origin. In this case and in many others it is convenient to take $t = 0$ and $s = 0$ at this point, but it is not always advisable to do so. We then choose a scale on each axis. In this case a convenient scale on the s-axis would be 1/4 inch to 20 feet and on the t-axis, 1/4 inch to a second. (It is here necessary to make the scale of s small in order to keep the graph on the paper.) Each value of t will then correspond to some value on the t-axis and similarly for s. Then we represent the pair of values $t = 1$, $s = 16$ by the point which is the inter-section of the vertical line through $t = 1$ on the t-axis and the horizontal line through $s = 16$ on the s-axis. Our table of values then will be represented by a set of four points as in Fig. 5:1, and we call these points the **graph** of the table of values.

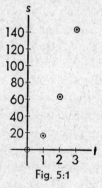

Fig. 5:1

The number pair (2,64), for example, is often written beside the point which is its graph and 2 and 64 are called the **coordinates** of the point. In order to avoid having to refer at length to "the point with $t = 2$ and $s = 64$" it is agreed that the first number of a pair of coordinates shall be

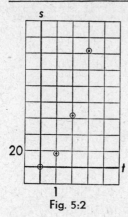

Fig. 5:2

measured horizontally and the second vertically. It is often convenient to use "coordinate paper," that is, paper which has rulings of horizontal and vertical lines. If we do that our graph above looks like Fig. 5:2.

As has been indicated, it sometimes is not convenient to have both variables take on the value 0 where the axes of the graph cross. Such a situation occurs with the following table of the temperature, T, on various days, D, of a patient with the measles.

Table 2 [1]

D	1	2	3	4	5	6	7	8
T	99.4	101.2	101.2	101.8	103.4	102.8	99	98.4

Here we can best show the graph by taking $T = 99$ at the origin and letting the vertical side of each square be .2°. We can have $D = 0$ at the origin and let the horizontal side of each square be 1.

Another less stringent agreement has to do with which set of data is to be laid out on the horizontal axis and which on the vertical. It is often not too important a matter, for the reasons for doing one thing or another may be largely esthetic or intuitive though none the less definite. Let us approach the matter with reference to our particular example. Here s and t vary and hence are called **variables**. The relationship we call a **function**. Whether or not there is a definite formula establishing the correspondence between one variable and the other, the idea of a correspondence itself rather implies that when we know the value of one, the value of the other is thereby determined. In that case we are thinking of the one as the so-called **independent variable** and the other as the **dependent variable**. It is cus-

[1] Gale and Watkeys, *Elementary Functions*, p. 34. Henry Holt & Co., New York, 1920.

tomary to assign the horizontal axis to the independent variable and the vertical axis to the dependent variable.

In many cases the choice of the independent variable is largely a matter of instinct with little logic behind it. In Table 1 we are apt to think of the distance as depending on the time and hence measure t along the horizontal axis. We might have trouble in persuading someone that the distance depended on the time rather than the time on the distance, but most persons would agree with our choice, especially since the time is recorded at regular intervals in the table. In Table 2 the temperature depends on the day. The situation is somewhat different in Table 3 where v is velocity in revolutions per minute and p is horsepower. Here a priori it might be difficult to decide whether the horsepower or revolutions per

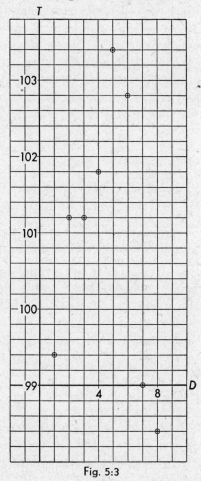

Fig. 5:3

minute were first determined. But the table itself gives us a clue. In the first place, the values of v are given

Table 3 [1]

v	1200	1500	1800	2100	2400	2700	3000	3300	3600
p	.94	1.06	1.15	1.15	1.07	.94	.75	.51	.31

[1] Griffin, F. L., *Introduction to Mathematical Analysis*, p. 20. Houghton Mifflin Co., Boston, 1921.

at regular intervals indicating that in the experiment from which the table arose, the values of v were first decided upon and then the horsepower calculated for these values. In the second place, any value of v in the table determines one accompanying value of p whereas in one case there are two values of p for the same v. Either one of these considerations would lead us to take v as the independent variable and graph it along the horizontal axis.

EXERCISES

1. In each of the following, state which variable you would consider the independent and which the dependent variable:

    a. The rate of interest on $1.00 compounded annually for 10 years and the amount accumulated.

    b. The height of a triangle of given base and its area.

    c. Variables a and b where the relationship is that a is 1 when b is a rational number and a is 0 when b is an irrational number.

    d. The number of students registered in Cornell and the price of bread in New Zealand.

2. Draw the graph of each of the following tables of values:

Table 4 [1]

E	0	6000	12000	18000	24000	30000	36000
P	30.0	23.8	19.0	15.0	11.8	9.5	7.5

    a. Table 4 where P is pressure in pounds and E is elevation in feet above sea level.

Table 5

t	0	10	15	35	45	60
s	0	$6\frac{2}{3}$	10	$23\frac{1}{3}$	30	40

    b. Table 5 where s is the distance in feet a body travels in t seconds.

Table 6 [2]

T	−20	−15	−10	−5	0	5	10	15	20	25	30	35	40
W	1	1.5	2.3	3.4	4.9	6.8	9.3	12.8	17.2	22.9	30.1	39.3	50.9

    c. A cubic meter of air at temperature $T°$ centigrade can hold W grams of water vapor.

[1] Griffin, F. L., *ibid.*, p. 9.
[2] *Ibid.*, p. 6.

Table 7

T	0	10	20	30	40	50	60
V	7500	7000	5500	3000	1300	600	300

d. It is estimated that T years hence a certain piece of property will be worth $\$V$.

Table 8 [1]

F	0	26	40	37	30	21	11	4	1
t	0	1	2	3	4	5	6	7	8

e. . . . where t seconds after starting a train, a locomotive exerts a force F tons more than the resisting forces.

Table 9

F	5′	6′	8′	10′	12′	15′	25′	50′	100′
W	2′4″	2′6″	2′9″	3′0″	3′2″	3′4″	3′8″	4′0″	4′2″

f. When working with a portrait attachment on a camera, focus at F feet to work at W feet and inches.

3. Introduction to the graphs of certain simple formulas.

If we know a formula connecting two variables we can construct our own table of values and find the graph of the relationship, that is, the graph of the formula or function. The formula and the picture augment each other in describing the relationship. Together they have the advantages of a talking picture over a photograph and a phonograph record.

Since various lines and curves have formulas or equations associated with them, it is possible to prove certain geometrical theorems by translating geometry into algebra, working with the algebra, and translating the results back into geometry. This interplay of the two subjects is called **analytic geometry** and is much more recent than the mere drawing of graphs, which dates back to three centuries before Christ. Descartes is generally given the credit for inventing analytic geometry, which he discussed in a book published in 1637, but the same ideas seemed to have occurred to

---

[1] Griffin, F. L., *ibid.*, p. 29.

others at slightly earlier dates. In this section, however, we shall content ourselves with merely drawing a few simple graphs of equations.

In this section and, in fact, usually when we are considering the graphs of equations or formulas, we choose the same scale on both axes and take as the origin the point whose coordinates are (0,0). If, then, a scale is fixed, we know that to every pair of positive real numbers a and b there corresponds a point a units to the right of the vertical axis and b units above the horizontal one. To the pair $(-a,b)$ corresponds a point a units to the left of the vertical axis and b units above the horizontal axis. The points corresponding to $(a,-b)$ and to $(-a, -b)$ are also shown in Fig. 5:4 below, always assuming that a and b are positive numbers. To every pair of real numbers there belongs just one point in the plane and to every point in the plane belongs just one pair. (Each pair is called an **ordered pair** since the order is important; that is, the point (a,b) is different from the point (b,a) unless $a = b$.) Notice that the axes divide the plane into four parts or **quadrants** which are customarily numbered as in the figure.

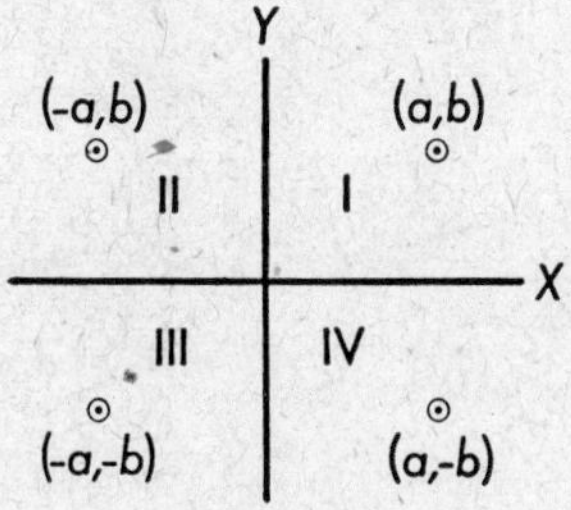

Fig. 5:4

The way we have chosen to set up a correspondence between the points of the plane and pairs of real numbers is only the most common of many ways to do this. The two axes need not be perpendicular nor even straight lines. Sometimes a point is located other than by its distance from two lines. But we shall remain loyal to our first love and shall not even cast a roving eye toward any other coordinate systems however beautiful they may seem to be. Unless something is said to the contrary, we shall abjectly take the x-axis to be horizontal and the y-axis vertical.

It should be pointed out that when we are dealing with an equation or formula we have more than a mere table of values. No matter how extensive a table of values we made, we should not be able to find all the pairs of values of the variables which satisfied the equation. So that when we have a formula (and as we shall see later under some other circumstances), it is desirable to connect the points found from the table of values by a smooth curve indicating that there are other points whose coordinates satisfy the equation but whose values we have not computed. We assume that all such points will lie in a smooth curve as, indeed, it can be proved is the case. Thus we consider the graph of an equation to have two important properties: *the coordinates of every point on the graph satisfy the equation and every pair of values of the variables satisfying the equation are the coordinates of a point on the graph.*

EXERCISES

1. Draw the graphs of the following equations:

   a. $x = 3$.   e. $y = -x$.
   b. $y = -5$.   f. $y = 2x$.
   c. $x = -7$.   g. $y = -3x$.
   d. $y = x$.   h. $-2y = x$.

2. Draw on a single set of axes the graphs of the following equations:

   a. $y = 2x$.   c. $y = 2x - 1$.
   b. $y = 2x + 3$ d. $y = 2x + 4$.

   If b is known merely to be a definite constant, what can you say about the graph of $y = 2x + b$?

3. Draw the graph of $y = -x + b$ for three different values of b.

4. Draw on one pair of axes the graphs of the equations:

$$y = x + 3, \quad y = 3x + 3, \quad y = -5x + 3, \quad y = -2x + 3.$$

   If a is known merely to be some constant, what can you say about the graph of $y = ax + 3$?

4. The equation of a straight line.

When we wish to describe numerically the steepness of a road we specify the **grade,** which is the distance it rises per unit distance measured along the road. In describing the steepness of a straight line, another ratio, called the **slope,** turns out to be more convenient. The slope of a line is the distance it rises from left to right per unit distance measured horizontally. In Fig. 5:5 the grades of the lines AB

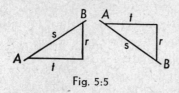

Fig. 5:5

are the ratio r/s and the slopes are the ratio r/t. Since any road not on a level may be uphill or downhill, depending on which way you are going, we customarily consider ourselves going from left to right in dealing with a straight line on a graph, or for that matter, with a curve. It is natural to consider the distance a line "rises" as *negative* when it is *descending* from left to right. Hence the first line in the figure has a positive grade and positive slope while the second line has a negative grade and a negative slope. It is permissible to speak of *the* slope of a line because the ratio does not depend on the points of observation. This may be

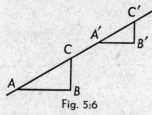

Fig. 5:6

seen to be true from Fig. 5:6 because ABC and $A'B'C'$ have corresponding sides parallel and hence are similar. It follows that $CB/AB = C'B'/A'B'$ and $CB/AC = C'B'/A'C'$. Furthermore, suppose P is some fixed point on a curve and it is true that the slope of the straight line PA is the same as that of PC no matter what points on the curve A and C are, then the curve is a straight line. This is true because we can consider A fixed for the moment; then no matter where C is on the curve it must be on the straight line PA, since there is only one line through P with the given slope. Our results then may be summarized in the following statement: **To say a curve has a fixed slope is equivalent to saying it is a straight line.**

We are now ready to deal with equations of straight lines. Suppose we have $y = 2x$. On this curve the value of y is always twice that of x. This means that if we take P to be the origin, the slope of PA is 2 no matter what point A is, provided only that its coordinates satisfy the equation. Hence the graph of $y = 2x$ is a straight line through the origin with the slope 2. Similarly **the graph of $y = mx$ is a straight line through the origin with the slope m.**

Suppose we have the equation $y = 2x + 3$. Here then, for each value of x, y is 3 units more than it is on the line $y = 2x$. We therefore have Fig. 5:7, and no matter what point A' is, $P'A'AP$ is a parallelogram and the slope of $P'A'$ is the same as that of PA, which is 2. Similarly, **the graph of $y = mx + b$ is a straight line through the point $(0,b)$ with the slope m.**

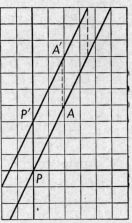

Fig. 5:7

Again, if we were given the equation $3x + 2y = 5$, we could solve it for y and have $y = -(3/2)x + 5/2$ which, by what we have shown above, is the equation of a straight line with slope $-3/2$ and passing through the point $5/2$. In fact, if $b \neq 0$, we can solve for y in the equation $ax + by + c = 0$ and have $y = -(a/b)x - c/b$ whose graph is a straight line of slope $-(a/b)$ and which passes through the point $(0, -c/b)$. If $b = 0$, the given equation reduces to $ax = -c$. Here if $a \neq 0$, we have $x = -c/a$ which is a straight line parallel to the y-axis and $-c/a$ units from it. If $a = 0$ our equation reduces to $c = 0$ which is satisfied by no values of x and y unless $c = 0$, in which case we really do not have much of an equation. Hence,

The graph of $ax + by + c = 0$ is a straight line provided a and b are not both zero. Also every straight line has such an equation which is therefore called a linear equation.

It is worth while comparing two special cases of the above equation. If $b \neq 0$ and $a = 0$ it reduces to $y = -c/b$ which may be written $y = 0 \cdot x - c/b$ and is a line of 0 slope, that is, a horizontal line. However, if $b = 0$ and $a \neq 0$ we have

$x = -c/a$ which is a vertical line. Its slope is not a number for if we refer to Fig. 5:5 we see that such a line would make $t = 0$ and the ratio r/t would not be a number. In other words, the slope of a vertical line is not a number but the slope of a horizontal line is a number, namely, zero.

Example: Draw the graph of $3x + 4y = 8$.

SOLUTION 1. Solve for y and have $y = -(3/4)x + 2$. Determine the point P with the coordinates $(0,2)$. The slope

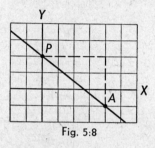

Fig. 5:8

is $-3/4$ and hence determine the point A which is 4 units to the right of P and 3 units down. PA is the line.

SOLUTION 2. We know the graph is a straight line. Hence any two points will determine the graph. When $y = 0$, $x = 8/3$, and when $x = 0$, $y = 2$. We find these two points on the graph and draw the line through them.

5. Nonlinear graphs.

The graph of any formula connecting two letters can be drawn simply by constructing a table of pairs of values satisfying the formula, determining the corresponding points on the graph, and connecting them in order by a smooth curve. For instance, suppose $y = x^2 - x$. We compute the following table of values:

x	0	1	2	3	−1	−2	−3
y	0	0	2	6	2	6	12

We know that no matter what x is we can find a value of y. Then, assuming that the points lie on a smooth curve, we have the graph given in Fig. 5:9. Notice that the lowest point in the curve is apparently $(\frac{1}{2}, -\frac{1}{4})$.

Suppose we wish the graph of $pv = 6$ where p stands for pressure in pounds and v is volume in cubic feet. Taking the point of view that pressure determines volume, we take

p to be the independent variable and measure it along the horizontal axis. Our table of values is

p	$\frac{1}{2}$	1	2	6	12
v	12	6	3	1	$\frac{1}{2}$

Since we cannot have a negative value for the volume or pressure, we assign no such values to the letters. Hence, letting one square represent two units for each axis, the

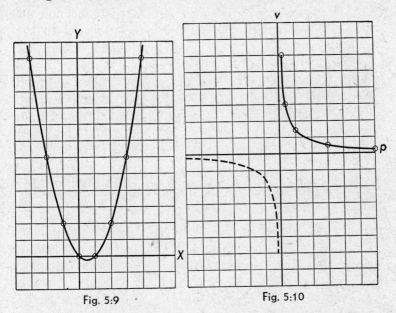

Fig. 5:9   Fig. 5:10

graph is only the solid curve in Fig. 5:10. However, if p and v had no physical meaning which required that they be positive, we should also have the dotted part of the graph.

As a final example, we draw the graph of $x^2 - y^2 = 9$. Solve this for y^2 and get $y^2 = x^2 - 9$, that is, $y = \sqrt{x^2 - 9}$ or $-\sqrt{x^2 - 9}$. If $x^2 < 9$ (that is, x^2 is less than 9), y is imaginary. Hence in our table of values occur no such values of x. Also, for each value of x there will be two values of y which are equal except for sign. Thus we list only nonnegative values of y and remember to include the negative values in the graph. The table of values is thus:

x	-3	3	4	-4	5	-5	7	-7
y	0	0	$\sqrt{7}$	$\sqrt{7}$	4	4	$\sqrt{40}$	$\sqrt{40}$

and the approximate value of $\sqrt{7}$ is 2.6 and of $\sqrt{40} = 2\sqrt{10}$ is 6.3.

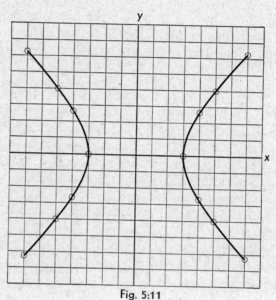

Fig. 5:11

EXERCISES

1. Draw the graphs of the following equations:
   a. $3x - 2y = 9$.
   b. $x = 5$.
   c. $y = x^2 - 3x + 2$.
   d. $y = x/(x-1)$.
   e. $x^2 - y^2 = 1$.
   f. $x^2 + y^2 = 25$.
   g. $s = 16\,t^2$ where s is the distance in feet a body falls in t seconds.
   h. $S = \dfrac{r^4 - 1}{r - 1}$ where S is the sum of a geometric progression having four terms, whose first term is 1 and whose fixed ratio is r.

2. The length plus the girth of any parcel to be sent by parcel post must not exceed 100 inches. Draw a graph showing the maximum allowable length for various values of the girth.

3. Draw a graph showing the amount of alcohol in various quantities of a 30% solution of alcohol and water.

4. Draw on one set of axes, the graphs of the lines

$$x + 2y = 3 \quad \text{and} \quad 3x - 2y = 1.$$

Find the coordinates of their point of intersection. What relationship do they have to the common solution of the pair of given equations? Under what circumstances will a pair of linear equations fail to have a common solution? Phrase your answer geometrically and algebraically.

5. Use Fig. 5:6 to show that the slope of a line through the points (x_1,y_1) and (x_2,y_2) is $(y_2 - y_1)/(x_2 - x_1)$.

6. Graphs of tables of observed values.

We have found two methods of obtaining tables of values: (1) by means of certain physical observations and (2) by the use of certain formulas. In the case of tables found from simple formulas experience leads us to expect that if we find certain values of the variables which satisfy the formula, graph the points corresponding to these values and draw a smooth curve through the points; then the coordinates of any point on the curve will satisfy the formula. So also in many types of physical tables, the recorded values of the variables are only a few of the totality of such values and we expect the graph of such a table to be a smooth curve within the limits of the physical problem.

For instance, in Table 5 the body will have traveled a certain distance in 5 seconds even though such an entry is not in the table. There certainly would be a value s for every value of t. If we assume that the body does not move in a fitful fashion we may consistently consider the graph to be a smooth curve and connect with a smooth curve the points corresponding to the entries in the table. Then on the above assumptions we can find from the graph the value of s when $t = 5$ and even the value of s when $t = 70$. These are guesses, of course, but they are intelligent guesses.

It is, as a matter of fact, on this assumption that nature is not too whimsical a lady that most of our orderly (and disorderly) lives are based. If I telephone a friend at twelve and at twelve-thirty only to find that the line is busy, I am apt to conclude that it was busy for the whole half hour; if the interval were 15 minutes, I should be even more sure of it. This is what might be called **between-sight** and our faith in it increases as the betweenness narrows. Foresight is more intriguing but not so reliable. Having observed the line busy at twelve and twelve-thirty I might not be so sure about my guess of not being able to reach my friend at twelve forty-five unless I heard on the party line the voice of the neighborhood gossip.

It is, of course, highly important to keep in mind the limitations imposed on the table of values by the physical situation. In Table 9, for instance, there would be no values for W when F less than zero nor, depending on the camera, for F greater than 100 feet. Furthermore, we shall soon have examples of tables of values whose graph is not a smooth curve. One such table would be a record of the number of eggs laid by a certain hen on the various days of her life.

EXERCISES

1. For which of the following tables of values or formulas should the graph be a smooth curve? Whenever such a curve is appropriate, draw it.

   a. to *f.* Tables 4 to 9.

   g. $S = n(n + 1)/2$, the formula for the sum $1 + 2 + 3 + \cdots + n$.

   h. $A = (1.01)^n$, the amount of $1.00 after n years at the interest rate of 1% compounded annually.

   i. The cost of sending a letter first class depending on its weight: 3 cents per ounce or fraction thereof.

   j. $s = 16\,t^2$ where s is the distance in feet fallen in t seconds.

2. Use the above graphs to estimate in Table 6 the value of W when $T = 12$ and the value of T which makes $W = 35$; in Table 7 the value of V when $T = 25$ and the value of T which makes $V = 6000$; in Table 8 the value of F when $t = 5.5$ and the value of t at which F has its greatest value; in Table 9 the value of W when F is 75 feet.

3. The cost ($C) of publishing a certain number (N) of pamphlets is computed by the formula $C = N/80 + 140$. Draw the graph of the formula for values of N from 1000 to 10,000. Use your graph to find the cost of publishing 2500 pamphlets. How many pamphlets may be published for $180? (Such a graph enables the printer to quote prices without computation. Graphs are used in this way in a variety of commercial enterprises.)

7. Finding a formula which fits or almost fits.

It often happens in a physical situation that we find a table of values by experiment or otherwise and seek a formula showing the relationship. This is desirable in that we can then use the formula as a means of "between-sight" (called **interpolation**) and foresight (called **extrapolation**) with greater accuracy than could be accomplished with the graph.

Of course, the first thing to do is to draw the graph of the points of the table and connect them by a smooth curve. If they seem to lie on a straight line or nearly so, we find the equation of the line through two points and see numerically how closely it approximates the others. Suppose we have the table:

x	1	4	5	7
y	3	9	11	14

The graph shows that the points lie approximately on a straight line. We call the line $y = ax + b$ and proceed to make it go through the points (1,3) and (4,9) by the following method. If it is to pass through the point (1,3) the equation must be satisfied when we substitute 1 for x and 3 for y. This gives us

$$3 = a + b.$$

Similarly we can make the line pass through (4,9) by requiring that

$$9 = 4a + b.$$

Subtract the first equation from the second to get

$$6 = 3a,$$
$$2 = a.$$

Putting this value into the first equation gives $b = 1$ and the line through the two points has the equation

$$y = 2x + 1.$$

When x is 5, this formula gives $y = 11$ and $x = 7$ gives $y = 15$. Thus the third point of the table lies on the line and the fourth is not far from it.

Next consider the table

x	1	5	8	11
y	2	22	58	112

The points here do not lie on a straight line. The next simplest equation is $y = ax^2 + bx + c$ whose graph is U-shaped and is called a **parabola**. Substituting the coordinates of the first three points into this equation yields the following:

(i) $2 = a + b + c,$

(ii) $22 = 25\,a + 5\,b + c,$

(iii) $58 = 64\,a + 8\,b + c.$

We get two equations in a and b by subtracting equation (i) from (ii) and subtracting equation (ii) from (iii). This yields

(iv) $20 = 24\,a + 4\,b,$

(v) $36 = 39\,a + 3\,b.$

Multiply equation (iv) by 3/4 to get

(vi) $15 = 18\,a + 3\,b$

and subtract equation (vi) from (v) to get

$$21 = 21\,a,$$
$$1 = a.$$

Substitute in equation (iv) to get

$$20 = 24 + 4\,b,$$

which gives $b = -1$. Then equation (i) gives

$$2 = 1 - 1 + c.$$

Hence $c = 2$ and the equation desired is

$$y = x^2 - x + 2.$$

To check the fourth entry in the table substitute $x = 11$ in the formula. This gives $y = 112$, which shows that all points of the table lie on the curve.

This process could be applied to find a curve of the third degree which passes through four given points but our attention shall be chiefly focused on the two equations above.

It sometimes happens that the values of the independent variable given in a table of values form an arithmetic progression. In that case one can, without drawing a graph, find which of the above two curves, if either, will fit the table. First, if the graph is a straight line, the values of the dependent variable will form an arithmetic progression when those of the independent variable are in an arithmetic progression, for the ratio of the difference between two values of the dependent variable and the difference between the two corresponding values of independent variable is the constant slope of the line. For instance, consider the following table where, in the line below the values of y, are listed the differences of successive values.

x	3	6	9	12	15
y	2	4	6	8	11
		2	2	2	3

The values of x form an arithmetic progression. The first four values of y also lie in an arithmetic progression. They thus lie on a straight line whose slope is 2/3, the ratio of the common differences of the y's and the common difference of the x's. The fifth point of the table lies close to but not on the line.

In general, let P, Q, and R be three points on a curve with the respective coordinates (a,A), (b,B), and (c,C) where a, b, and c are in an arithmetic progression, that is, $b - a = c - b = d$. The slope of PQ is $(B - A)/(b - a) = (B - A)/d$ and the slope of QR is $(C - B)/(c - b) = (C - B)/d$. Thus, if A, B, and C are in an arithmetic progression, $C - B = B - A$, the slope of PQ is equal to the slope of QR and hence P, Q, and R lie on a straight line. On the other hand, if PQ and QR have equal slopes, $C - B = B - A$ and A, B, and C are in an arithmetic progression.

It turns out to be true that an extension of this phenomenon applies to the other case considered above. In the table below we list the differences of the values of y — the so-called **first differences** — and the differences of the first differences — the so-called **second differences**.

x	1	3	5	7	9
y	2	8	22	44	74
First differences		6	14	22	30
Second differences			8	8	8

Notice that here the first differences form an arithmetic progression. Whenever this happens it is true that the formula $y = ax^2 + bx + c$ will fit the table. By the methods above used we could find the formula to be $y = x^2 - x + 2$.

We now prove that if in any table the values of x as well as the first differences form an arithmetic progression, then a formula $y = ax^2 + bx + c$, for properly chosen values of a, b, and c is satisfied by all pairs of values of the table.

First we show that if (r,s), (t,u), and (v,w) are three points and if no two of r, t, v are equal (that is, if no two of the points be on the same vertical line), then a, b, and c may be determined so that the curve $y = ax^2 + bx + c$ goes through the three points. Furthermore, a, b, and c are completely determined by the three points (just as the equation of a line is determined by two points). To show this, notice that since the points are to satisfy the equation, we may substitute their coordinates for x and y in the equation and get

$$s = ar^2 + br + c$$
$$u = at^2 + bt + c$$
$$w = av^2 + bv + c$$

Subtract the second equation from the first and the third from the second to get the two equations

$$s - u = a(r^2 - t^2) + b(r - t)$$
$$u - w = a(t^2 - v^2) + b(t - v)$$

To solve for a in the last two equations we can multiply the first by $t - v$ and the second by $r - t$ and subtract. Since we are not interested in the solution itself, but only in whether or not there is a solution, we need only be concerned with the right side of the resulting equation, which is

$$a(r^2-t^2)(t-v) - a(t^2-v^2)(r-t) = a[(r^2-t^2)(t-v) - (t^2-v^2)(r-t)].$$

We can solve for a provided that which multiplies it is not zero. But

$$(r^2 - t^2)(t - v) - (t^2 - v^2)(r - t) = (r - t)(t - v)(r + t - t - v)$$
$$= (r - t)(t - v)(r - v)$$

which cannot be zero since no two of r, t, and v are equal. Hence we can solve for a. With a known, the equation $s - u = a(r^2 - t^2) + b(r - t)$ can be solved for b since $r - t \neq 0$ and, with b known, the first equation can be solved for c. Notice, incidentally, that if $r - t = 0$, then $y = ax^2 + bx + c$ would imply $s - u = 0$ and two of the given points would coincide. What does this show?

Now suppose we have a table of values, Table A, in which the values of x as well as the first differences of the y's form an arithmetic progression. We have just shown that we can determine a, b, and c so that $y = ax^2 + bx + c$ is satisfied by the first three pairs of values of the given table. If, then, we form a second table, B, of values of y computed from the formula, certainly the first three entries in both tables will be the same. If we can show that the first differences in the Table B form an arithmetic progression, it must be the same arithmetic progression as in Table A (since the first two numbers are the same) and hence the values of y must be the same in both tables. Hence we need now to show that if the values of x are in arithmetic progression the first differences of the values of y are in arithmetic progression where $y = ax^2 + bx + c$. Now y is the sum of ax^2, bx, and c. Hence the first differences of the y's will be the sum of the first differences of ax^2, bx, and c. We form separate tables of the first differences of each of these:

x	s	$s + d = t$	$s + 2d = u$	$s + 3d = v$
c	c	c	c	c

First differences $\qquad$ 0 $\qquad$ 0 $\qquad$ 0

bx	bs	$bs + bd$	$bs + 2bd$	$bs + 3bd$

First differences $\qquad$ bd $\qquad$ bd $\qquad$ bd

ax^2	as^2	at^2	au^2	av^2

First differences $\qquad$ $at^2 - as^2$ $\qquad$ $au^2 - at^2$ $\qquad$ $av^2 - au^2$

We have previously seen that the first differences of the squares 1 4 9 16 25 ⋯ form an arithmetic progression. The same would be true if we took every third number: 1 16 49 etc., or every dth number. Thus the first differences of ax^2 in the table above are a times the numbers of an arithmetic progression, that is, form themselves an arithmetic progression; adding $bd + 0$ to each term of this arithmetic progression results in another arithmetic progression. Thus the sums of the first differences in the table above form an arithmetic progression and we have proved what we set out to prove.

It can be shown along similar lines that an equation of the form $y = ax^3 + bx^2 + cx + d$ can be found passing through any four points no two on the same vertical line and that if, in any table of values, the second differences form an arithmetic progression, an equation of this form will be satisfied by the points of the table. Similar results hold for equations of higher degrees.

EXERCISES

1. Find the equations of the lines through
   a. $(3,-2)$ and $(5,1)$
   b. $(6,4)$ and $(-2,-3)$
   c. $(-2,1)$ and $(-2,5)$
   d. $(0,0)$ and $(-1,-8)$
   e. $(1,-5)$ and $(3,-5)$

2. Show that there is a line $y = ax + b$ passing through the points (r,s), (t,u) if $r \neq t$. What happens if $r = t$?

3. Table 10 gives the weight (W grams) of potassium bromide which will dissolve in 100 grams of water at various temperatures. Draw the graphs of the points of the table. Find and check the equation for the line on which all points of the graph lie. From the graph of the line find W when $T = 23$ and check by using the formula.

Table 10 [1]

T	0	20	40	60	80
W	54	64	74	84	94

4. Table 11 gives the volume of a certain quantity (V cubic centimeters) of gas at several temperatures ($T°$). Draw the graph of the points, find and check the formula giving V in terms of T. Find V when $T = 50$.

Table 11 [2]

T	-33	-6	12	27	42
V	160	178	190	200	210

5. Table 12 gives the average weights (W pounds and M pounds) of women and men for various heights (h inches). Draw the graphs of W and M as functions of h. In each case find the equations of lines on which the points very nearly lie.

Table 12 [3]

h	61	63	65	67	69	71
W	118	126	134	142	150	158
M	124	132	140	148	157	166

6. Find the equation of a line which very nearly goes through all the points of the graph of the percentage of alcohol required for various temperatures as given in section 9 of Chapter IV. What will be the percentage required for a temperature of $-5°$?

[1] Griffin, F. L., *Introduction to Mathematical Analysis*, p. 52.
[2] *Loc. cit.*
[3] Griffin, F. L., *ibid.*, p. 53.

7. We have the following table of values of S for various values of n where $S = 1 + 2 + 3 + \cdots + n$.

Table 13

n	1	2	3	4	5
S	1	3	6	10	15

By the methods of this section, find a formula for S in terms of n.

8. Find a formula which fits or almost fits the following table:

Table 14

t	2	5	8	11	14
s	0	15	48	99	169

9. Find the formula which fits or almost fits the table:

Table 15

x	0	2	4	6	8
y	0	10	45	102	184

10. A contractor agreed to build a breakwater for $250,000. After spending $66,600 as indicated in Table 16, he threw up the job without receiving any pay because he estimated that the cost would increase according to the same law and that it would require six months more to complete the work. Find a formula approximating the table and use it to estimate what the loss would be had he finished the breakwater.

Table 16

Time in months	0	1	2	3	4
Investment in thousands	10	12.2	22.4	40.6	66.6

11. Find a formula of the form $S = an^3 + bn^2 + cn + d$ which fits the following table of values of S in terms of n where

$$S = 1^2 + 2^2 + 3^2 + 4^2 + \cdots + n^2:$$

Table 17

n	1	2	3	4	5
S	1	5	14	30	55

12. Find a formula of the form S in Exercise 11 fitting the following table of values:

Table 18

n	1	2	3	4	5
S	1	4	10	20	35

These are the so-called pyramidal numbers — the number of cannon balls in various triangular pyramids. For any n, the value of S is the sum of the first n values of S in Table 13.

13. Show that no matter how far one continues to take differences of differences in Table 19, one will not get an arithmetic progression. Reasoning by analogy from your experience above, what conclusions would you draw from this?

Table 19

n	1	2	3	4	5	6	7	8
S	1	1	2	3	5	8	13	21

Each S is the sum of the two previous ones. These values of S form the so-called **Fibbonacci series.**

8. Frequency tables.

Generally speaking, the graphs we have been considering have involved infinitely many points — usually a curve — and this has been acknowledged by joining by a curve the scattered points corresponding to the pairs in the table of values. The justification for this procedure is that we were interested mainly in tables of values which arose either from measurements of speed, temperature, time, etc., or from formulas which gave a significant result whenever any real number (within limits) was substituted for the independent variable. The entries in our table of values were merely samples from a set of infinitely many values and which we computed or observed to give us an idea of the trend.

In this section, we wish to examine certain special tables and their graphs in which, either in the nature of the case or due to some definite agreement, the dependent variable does not have a value for each real value of the independent variable between any two limits. For instance, in the formula $S = n(n + 1)/2$ for the sum of the first n integers, n has only positive integral values. We can, to be sure, find a value of S from the formula when $n = 1/2$ or $n = -1$, yet there would be no sense in speaking of "the sum of the first half integer" or "the first minus one integers."

In many statistical tables, there are a few entries and no

more. Suppose it is required to measure the heights of 1,000,000 men, shall we say, in the army. The heights will range from, perhaps, 62 inches to 77 inches. The height of a given man can be any real number between these limits. The processes of measurement used in such circumstances are not accurate enough, however, to distinguish differences in height less than perhaps 1/32 of an inch. In fact, the usual procedure is to group together all those whose heights fall between 62 and 63 inches, 63 and 64 inches, $\cdots$, and between 76 and 77 inches, where each interval includes the lower but not the upper measurement; to these groups are assigned the heights 62.5, 63.5, $\cdots$, 76.5 inches respectively. With each of these groups is associated the number of men in the group. Thus we get a table of values like the following:

h	62.5	63.5	$\cdots$	76.5
n	n_1	n_2	$\cdots$	n_{15}

where $n_1 + n_2 + \cdots + n_{15} = 1,000,000$. Notice that there is certainly no value of n for values of h between 62.5 and 63.5, for instance, and hence it would be *very misleading* to connect by a curve two points of the table of values. Similar remarks could be made for the following.

Example. One hundred shells were fired at a target and the results tabulated according to their striking F feet short of (when a minus sign is used) or beyond (when a plus sign is used) the target. The results are then arranged into ten groups so that in the table, $N = 25$ for $F = -5$ means that 25 shells fell between -10 and 0 feet from the target, etc. The table is

Table 20 [1]

F	-35	-25	-15	-5	$+5$	$+15$	$+25$	$+35$
N	2	7	16	25	25	16	7	2

Tables such as these which record the frequency with which a certain event occurs are called **frequency tables**. We have seen that they call for a different kind of graph.

[1] Griffin, F. L., *ibid.*, p. 455.

Sometimes successive points are joined by straight lines forming what is called a **frequency polygon.** Here we use what is called a **histogram** and to show what we mean by such a form of graph we construct it for Table 20. We begin by constructing the graph of the *points* of the table in the usual way. Then, since +5 covers the range from 0 to 10 we construct a rectangle whose base is the line from

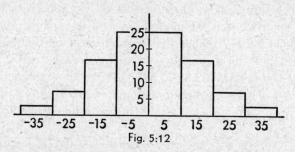

Fig. 5:12

points 0 to 10 on the F-axis (horizontal) and whose altitude is 25, the corresponding value of N. So we proceed to form Fig. 5:12.

The histogram form of graph has the definite advantage that it shows the range which a specific number stands for; e.g., 15 is the number used for the range 10 to 20. Even in such cases as Table 23 below in which no range of values is involved, the histogram indicates lack of smoothness of the data.

EXERCISES

1. In Table 21, N denotes the number of telephone calls, during a certain period of the day, which last T seconds. (The calls have been grouped, of course, as the heights were in the examples above.)

Table 21 [1]

T	50	150	250	350	450	550	650	750	850	950
N	1	28	88	180	247	260	133	42	11	5

Construct a histogram for this table.

2. Draw the histogram for the following mortality table, Table 22,

[1] Kenney, *Mathematics of Statistics*, p. 18. D. Van Nostrand Co., New York, 1939.

Table 22

A	10	15	20	25	30	35	40	45		
L	100,000	96,285	92,637	89,032	85,441	81,822	78,106	74,173		

A	50	55	60	65	70	75	80	85	90	95
L	69,804	64,563	57,917	49,341	38,569	26,237	14,474	5485	847	3

where L denotes the number of survivors at age A of a group of 100,000 alive at age ten.

3. In Table 22 let all people dying between their tenth and fifteenth birthdays be regarded as dying at age 12.5 and so on. For instance, we should say that the last three people die at age 97.5. Let D denote the number dying at age a, thus defined, and complete the following table:

a	12.5	17.5	22.5	$\cdots$	87.5	92.5	97.5
D							

Construct a histogram for this table. What connection would there be between it and the histogram for the mortality table?

4. A set of 10 coins was tossed 1000 times, and the number H of heads appearing each time was recorded. Letting N denote the number of times H heads appeared, the following table of observations resulted:

Table 23

H	0	1	2	3	4	5	6	7	8	9	10
N	2	10	41	119	206	246	204	115	44	11	2

Construct the histogram for this table.

5. The marks of a class of ten students are as follows:

$$60, 63, 65, 68, 72, 73, 75, 85, 86, 90.$$

Exhibit these marks graphically, first giving the marks just as they are; second, grouping them in four ranges: 60–69, 70–79, 80–89, 90–100. In each case use the form of the graph which you consider most appropriate.

6. Name the form of graph most suitable for each of the following tables of values:

a. Table 24, where A is the number of apples having S seeds.

Table 24

S	4	5	6	7	8	9
A	9	4	14	21	24	23

b. Table 25 where v is the velocity in feet per second of a falling body after t seconds.

Table 25

v	11	27	43	59
t	1	2	3	4

c. The principal unpaid at the end of each year under Plan A and Plan B of example 5, pp. 128–130.

d. Table 26, where I is the inaccuracy in inches allowed on a ring gauge when measuring distances of d inches, the lower measurement in each interval excluded.

Table 26 [1]

d	.029 to .825	.825 to 1.510	1.510 to 2.510	2.510 to 4.510
I	.00020	.00024	.00032	.00040

d	4.510 to 6.510	6.510 to 9.010	9.010 to 12.010
I	.00050	.00064	.00080

e. Table 27, where if a man pays $100 a year beginning at age 40 until age A he can receive, beginning at age A, M dollars per month for the rest of his life.

Table 27 [2]

A	50	55	60	65	70
M	4.97	9.22	15.31	24.27	37.70

f. The graph of the table for Exercise 3, in section 6 of Chapter IV.

9. Averages.

One of the most maltreated terms in the English language is the word "average." The "average man" thinks or does thus and so, weighs such and such, and therefore we are expected to think, do, and weigh likewise. He is set up as an example for all of us to emulate and woe be it to him who is branded as being "below average"! The average is one of the greatest comforters of mankind for we have the feeling that if we can only take the average everything will come out all right. The average of several inaccurate measurements somehow ought to be better than any one of them

[1] International Business Machines Corp., *Precision Measurement in the Metal Working Industry*, Chap. IV, p. 6. Reprint for the University of the State of New York, 1941.

[2] Harwood and Francis, *Life Insurance from the Buyers' Point of View*, p. 145. American Institute for Economic Research, Cambridge, Mass., 1940.

though we do hesitate to use this panacea with our results of several additions of a given column of figures.

But the general use of this idea of average, however loosely it is employed, is a sign of a very definite and legitimate need to characterize a whole set of data by means of a single one. As the eminent British statistician, R. A. Fisher, puts it: "A quantity of data which by its mere bulk may be incapable of entering the mind is to be replaced by relatively few quantities which shall adequately represent the whole, or which, in other words, shall contain as much as possible, ideally the whole, of the relevant information contained in the original data." While an average does not give a very accurate idea of a trend in a set of data, it does describe the situation "in a nutshell."

We shall deal with five kinds of averages, defining them in terms of two tables:

Table 28

G	1	4	6	8
N	3	2	1	4

Table 29

x	x_1	x_2	$\cdots$	x_n
f	f_1	f_2	$\cdots$	f_n

where Table 28 is the record of the grades of a certain class of ten students (three have grade 1, two the grade 4, etc.) and Table 29 is a general frequency table in which x_1 occurs f_1 times, x_2 occurs f_2 times, etc. We say that f_1 is the frequency of the observation x_1.

The **mode** is defined to be the measurement which has the greatest frequency. In Table 28 it is 8 since more students have that grade than any other. In Table 29 it is the x corresponding to the greatest f. There may be several modes. For instance, if, in Table 28, four students had grade 1, four grade 8, and less than four each of the other grades, there would be two modes: 1 and 8.

The **median** is the middle observation or halfway between

the two middle ones. This can perhaps be more easily understood by merely listing all the grades in order for Table 28 as in

(A)   1 1 1 4 4 6 8 8 8 8.

There are ten grades in all, half of ten is 5, and hence the "middle" grade would be 5, which is halfway from 4 to 6. In case there were eleven grades, the median would be that one which occurs sixth from either end. It, of course, would be laborious to write every table in form (A) and hence the usual process for finding the median would be this: find the total number of observations, that is, the sum of all the frequencies, which we may call F. If F is odd, divide $F + 1$ by 2 and the median will be the observation corresponding to the $(F + 1)/2$th frequency counting from either end of the table. If F is even, the median will be the observation halfway between the *two* middle observations, namely between the $F/2$th and the $1 + F/2$th observations.

The **arithmetic mean,** which is usually what is meant by the term "average," is the sum of all the observations divided by the total number of frequencies. Using form (A) of Table 28 we have

$$\frac{1 + 1 + 1 + 4 + 4 + 6 + 8 + 8 + 8 + 8}{10}.$$

This is obtained with less writing from Table 28 directly as

$$\frac{3 \cdot 1 + 2 \cdot 4 + 1 \cdot 6 + 4 \cdot 8}{3 + 2 + 1 + 4} = 4.9.$$

For Table 29, the arithmetic mean would be

$$\frac{f_1 x_1 + f_2 x_2 + \cdots + f_n x_n}{f_1 + f_2 + \cdots + f_n}.$$

The **geometric mean** is the Fth root of the product of the grades or observations. Using form (A) we would have

$$\sqrt[10]{1 \cdot 1 \cdot 1 \cdot 4 \cdot 4 \cdot 6 \cdot 8 \cdot 8 \cdot 8 \cdot 8}.$$

Using Table 28 directly we may represent this with less writing as

$$\sqrt[10]{1^3 \cdot 4^2 \cdot 6 \cdot 8^4} = 3.6, \text{ approximately.}$$

For Table 29, it would be

$$\sqrt[F]{x_1^{f_1} \cdot x_2^{f_2} \cdot \cdots \cdot x_n^{f_n}}$$

where F, you recall, is the sum of the values of f. This mean cannot easily be computed except by means of logarithms, the use of which is a laborsaving device beyond the scope of this book.

The **harmonic mean** is the total number of frequencies, F, divided by the sum of the reciprocals of the observations. Using form (A) we have

$$\frac{10}{\frac{1}{1} + \frac{1}{1} + \frac{1}{1} + \frac{1}{4} + \frac{1}{4} + \frac{1}{6} + \frac{1}{8} + \frac{1}{8} + \frac{1}{8} + \frac{1}{8}}.$$

Using Table 28 directly, this amounts to

$$\frac{10}{\frac{3}{1} + \frac{2}{4} + \frac{1}{6} + \frac{4}{8}} = 2.4.$$

For Table 29 it would be

$$\frac{F}{\frac{f_1}{x_1} + \frac{f_2}{x_2} + \cdots + \frac{f_n}{x_n}}.$$

In terms of the histogram of a distribution, the mode is that x (or those x's) at the center (or centers) of the base (or bases) of the highest rectangle (or rectangles). The median is that number on the horizontal axis with the property that a vertical line through it bisects the total area of all the rectangles. (Why?) The arithmetic mean is that number on the horizontal axis with the property that a vertical wire through it will support in equilibrium a sheet of cardboard or other material of homogeneous weight cut in the form of the histogram. (That is, the histogram can be balanced on a single point of support at the point which represents the frequency for the arithmetic mean.) Notice that the above discussion would apply only to averages associated with a frequency table since for computation, except for the mode, it was necessary to find the total number of frequencies which, by necessity, would need to be a finite number. However, even for the graph of a continuously varying function, such as speed in terms of time, one can draw the graph and define the arithmetic mean and median in terms of the area under the curve. This, however, requires more advanced methods than we can use in this book.

While the first three averages are more or less catholic in their application, the last two are used for special purposes. The geometric mean is the proper average to use when ratios of values are important, because of the property that the ratio of the geometric means of two sets of corresponding measurements is equal to the geometric means of their ratios. Suppose, for instance, that one box has the edges a, b, and c and another has edges A, B, and C. The ratio of the geometric means is

$$\frac{\sqrt[3]{abc}}{\sqrt[3]{ABC}}.$$

On the other hand, the ratios of the corresponding sides are a/A, b/B, c/C whose geometric mean is

$$\sqrt[3]{\frac{a}{A} \cdot \frac{b}{B} \cdot \frac{c}{C}} = \sqrt[3]{\frac{abc}{ABC}} = \frac{\sqrt[3]{abc}}{\sqrt[3]{ABC}}.$$

In comparing the relative sizes of the boxes, it is the ratios which are important. This same average would be appropriate if a, b, c were the respective prices of three certain commodities in a certain year, and A, B, C their prices another year, for it is the ratios of corresponding prices that is significant in describing the price level.

The name **harmonic mean** stems from the fact that if the musical interval from note a to b is the same as from b to c, the length of the string producing b is the harmonic mean of the lengths for a and c. Another good illustration of the use of this mean is afforded by problem 9 of section 10 in Chapter IV. We can see what is happening better if we substitute letters for the given numbers and say that the "out journey" was at the rate of a miles per hour and the return journey at b miles per hour. Let d be the one-way distance. The total time taken is then $c/a + d/b$ and the distance traveled is $2\,d$. The average rate of the trip is the distance divided by the time and hence is

$$\frac{2\,d}{\dfrac{d}{a} + \dfrac{d}{b}} = \frac{2\,d}{d\!\left(\dfrac{1}{a} + \dfrac{1}{b}\right)} = \frac{2}{\dfrac{1}{a} + \dfrac{1}{b}}$$

which is the harmonic mean of these two rates.

EXERCISES

1. Find the median and mode of Tables 21, 22, and 23.

2. After grouping, the weights of 1000 eight-year-old girls yielded the following frequency table:

Table 30 [1]

W (lb)	29.5	33.5	37.5	41.5	45.5	49.5	53.5	57.5	61.5	65.5
f	1	14	56	172	245	263	156	67	23	3

Find the mode, median, and the arithmetic mean for this table.

3. In a certain test, the 12 girls in a class made a mean grade of 72 and the 15 boys made a mean grade of 70. Find the mean grade of the entire class. (By "mean" is meant the arithmetic mean.)

4. Compare the arithmetic mean, mode, and median for each of the following tables.

Table 31

x	0	1	2	3	4	5	6
f	2	7	13	16	13	7	2

Table 32

x	0	1	2	3	4	5	6
f	1	2	19	16	1	1	20

On the basis of your results can you make any guesses as to the behavior of the three averages for various frequency tables?

5. Concoct a set of grades of 10 students, 9 of whom have grades below the arithmetic mean.

6. Would it be possible for the arithmetic mean of the grades of a class of ten to be above 70 and yet seven of the class to get grades below 60?

[1] Kenney, *ibid.*, p. 18.

7. In baseball circles one speaks of a player's "batting average." Is this one of the averages we have discussed above? If so, which one?

8. How would you give a numerical evaluation of the "average man" and "average diet"?

9. Find the geometric mean of the four numbers: 2, 6, 9, 12 and compare it with the arithmetic and harmonic means.

10. A man drives at the average rate of 15 miles per hour through the towns. On a certain 50-mile journey he passes through three towns each a mile long. If he is to make the journey in an hour, what must be his average rate in miles per hour on the open road (that is, outside the towns)? What would be your answer if there were t towns each a mile long?

11. In Pennsylvania the speed limit in the towns is 15 miles per hour and outside the towns 50 miles per hour. Mr. Thims driving from city A to city B, finds that one-tenth of the distance is through the towns. Can he legally average 40 miles an hour for the trip? Does your result depend on the distance between A and B? Explain your answers.

12. The population of a certain city is doubled in one decade and quadrupled in the next. What would be the most appropriate average to use in answering the following question: what was its average ratio of gain per decade? Explain the reasons for your answer.

13. Prove: to say that b is the harmonic mean of a and c is equivalent to saying that $1/a$, $1/b$, $1/c$ form an arithmetic progression.

14. A sequence of numbers is called an **harmonic progression** if each number of the sequence is the harmonic mean of the number preceding it and the number following it. Prove that

$$1, 1/2, 1/3, 1/4, \cdots$$

is an harmonic progression.

15. Prove that any harmonic progression can be written in the form

$$1/a_1, 1/a_2, 1/a_3, \cdots$$

where $a_1, a_2, a_3, \cdots$ form an arithmetic progression.

* 16. Prove that the geometric mean of two unequal positive numbers is always less than the arithmetic mean and greater than the harmonic mean. How do the means of a pair of equal numbers compare?

* 17. Fill in the frequencies in the frequency table

x	-1	0	1
f	f_1	f_2	f_3

so that the magnitudes of median, mode, and arithmetic mean have various possible numerical orders. For instance, if the values of f are 2, 1, and 3, the median is 1/2, the mode is 1, and the mean is 1/6, giving an example of a case in which the mean is the least, the median next, and the mode greatest. Some orders are impossible to attain.

10. Topics for further study.

1. Analytic geometry: See reference **29,** Chap. 7.

2. Curve-fitting (see section 7): See reference **10, pp.** 101–104.

3. Averages: See reference **17,** pp. 388–398.

4. Methods of summarizing data: See reference **17, pp.** 382–431; **31, pp.** 378–382.

CHAPTER VI

Permutations, Combinations, and Probability

---

1. Routes and permutations.

In the last chapter the frequency tables which we considered were obtained from experiments or from observing things happen. There are other frequencies of a rather different character which can be calculated by means of a kind of mental experiment where we see in our mind's eye in how many ways certain things occur or happen.

Suppose, for instance, that there are two roads connecting Ithaca, I, and Cortland, C, and three roads connecting Cortland and Syracuse, S. A driver starting from Ithaca to

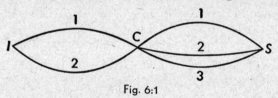

Fig. 6:1

Syracuse may be interested to know in how many different ways he can plan his route. A look at the diagram below answers the question almost immediately. The possible routes to Syracuse are

$$11, 12, 13, 21, 22, 23$$

and one sees that there are six of them. Notice that one obtains the answer by *multiplying* 2 (the number of roads connecting Ithaca and Cortland) by 3 (the number of roads connecting Cortland and Syracuse).

Another example of a problem leading to the question "In how many ways can it be done?" is the following:

Someone wants to distribute a knife, a lighter, and a pen among three of his friends: John, Henry, and Robert. In how many different ways can he do it?

If one decides to give the knife to John, the lighter to Robert, and the pen to Henry, we shall denote it symbolically by

$$k \quad l \quad p$$
$$J \quad R \quad H$$

The letters k, l, p are abbreviations of "knife," "lighter," "pen" and J, R, H the initials of the friends. It is easy to see now that the answer to our problem is 6, since the following possibilities are the only ones

k	l	p		k	l	p		k	l	p
J	H	R		J	R	H		H	J	R

k	l	p		k	l	p		k	l	p
H	R	J		R	J	H		R	H	J

If two sequences of symbols differ only in order we say that they are **permutations** of each other. Thus k, l, p is a permutation of p, k, l (and vice versa); 3, 1, 2 is a permutation of 2, 1, 3; but a, 1, 2 is not a permutation of 1, 6, 2; and 1, 1, 3 is not a permutation of 1, 3, 3. (Cf. Chap. II, section 15.)

The problem we just solved was equivalent to finding how many different permutations of three symbols there are. It is obvious that the nature of the symbols is immaterial and that just as there are six permutations of 3, 1, 2, so there are six permutations of k, l, p or *, /, 1.

Now we shall find the number of permutations of four symbols. To simplify the writing we shall assume that our symbols are the numbers 1, 2, 3, and 4. Let us consider first of all those permutations which begin with 1. Every such permutation can be written in the form

$$1 \quad a \quad b \quad c$$

where a, b, c is a certain permutation of the three remaining symbols: 2, 3, and 4. As we know, there are six permutations of three symbols and there are, therefore, six permutations of four symbols which begin with 1. The same reasoning leads to the immediate result that there are six permutations of four symbols which begin with 2, six which begin with 3, and six which begin with 4. Hence the total number of permutations of four symbols is

$$6 + 6 + 6 + 6 = 4 \cdot 6 = 24.$$

Exactly the same kind of reasoning leads to the result that there are

$$24 + 24 + 24 + 24 + 24 = 5 \cdot 24 = 120$$

permutations of five symbols. Notice that $120 = 1 \cdot 2 \cdot 3 \cdot 4 \cdot 5$.

Even though the answers to the route problem and the knife-lighter-pen problem are both 6, the problems themselves are very different, for no permutations enter into the former. These two problems have been introduced in this section to emphasize the necessity of treating each problem on its own merits without any attempt at classification.

EXERCISES

1. There are 3 roads connecting A and B, and 4 roads connecting B and C. In how many different ways can one reach C by way of B starting from A?

2. There are 2 roads from A to B, 3 roads from B to C, and 3 roads from C to D. In how many different ways can one reach D from A?

3. Prove that there are 720 permutations of 6 symbols.

4. How many different 6-digit numbers can be formed by using the digits 1, 2, 3, 4, 5, 6, no digit being used twice?

5. Prove that there are $n!$ permutations of n symbols where $n!$ means $1 \cdot 2 \cdot 3 \cdots n$.

6. How many different 6-digit numbers can be formed by using the digits 0, 1, 2, 3, 4, 5, no digit being used twice?

7. How many of the numbers in Exercise 6 are divisible by 5?

SOLUTION: A number is divisible by 5 if and only if the last digit is either 5 or 0. The numbers are therefore of the form

$$a\,b\,c\,d\,e\,5 \quad or \quad a\,b\,c\,d\,e\,0.$$

There are 120 numbers of the second kind, for the number of permutations of 5 symbols is 120; and $120 - 24 = 96$ numbers of the first kind, for we must exclude the 24 permutations of a, b, c, d, e which begin with 0. Hence the answer is $120 + 96 = 216$.

8. How many of the numbers of Exercise 4 are even? Of Exercise 6?

9. How many of the numbers of Exercise 4 are divisible by 3? How many are divisible by 9? By 6?

10. How many different 6-digit numbers can be formed containing only the digits 1, 2, 3, 4, 5, 6, in which any digit may occur any number of times?

2. Combinations.

Three problems somewhat different from those which we have considered above are the following:

Problem 1. An examination contains five questions but the student is supposed to answer only two of them. In how many different ways can he make his choice? Enumerating the questions one can see that the following are the only possibilities:

$$(1, 2) \quad (2, 3) \quad (3, 4) \quad (4, 5)$$
$$(1, 3) \quad (2, 4) \quad (3, 5)$$
$$(1, 4) \quad (2, 5)$$
$$(1, 5)$$

where $(2, 5)$, for instance, means that the student decides to answer the questions numbered 2 and 5. The answer to the problem is 10.

Problem 2. In how many different ways can a student choose three out of five questions? Again the answer is 10, for choosing three questions to be answered is equivalent to choosing two questions which the student will leave unanswered; we can, therefore, use the same table as above with the understanding that $(2, 5)$, for instance, means that the student leaves *unanswered* the questions numbered 2 and 5.

Problem 3. A student is to read two out of five of Shakespeare's tragedies. In how many different ways can he make up his mind? The answer here is also 10 and the problem differs from the first only by the nature of things involved (tragedies instead of questions).

A sequence of symbols written in an arbitrary order is called a **combination.** For example, a, b, c is a combination, and b, c, a is the same combination because the same symbols are used, though in a different order. But a, b, c and b, z, d are different combinations because they involve different symbols. If $m \leqq n$ we shall denote by $_nC_m$ the number of different combinations of m symbols taken from a given set of n symbols, that is, the number of combinations of n things taken m at a time. In this notation, the answer to the first problem is

$$_5C_2 = 10$$

and the answer to the second

$$_5C_3 = {_5C_2} = 10.$$

We have found the number of certain combinations above by writing down all of them, but this procedure will obviously fail in case we want to find, for instance, $_{100}C_{81}$. It would take considerably longer than a lifetime for one man to write down all these combinations. (By using formula (3) below let the interested reader estimate how long it would take.) There is, however, a way of evaluating the symbol $_nC_m$ whatever positive integers m and n may be, provided only $m \leqq n$. We shall illustrate the method on the familiar problem of finding $_5C_3$. Let the symbols at our disposal be a, b, c, d, and e. We first divide all the combinations into two classes. The first class will contain the combinations in which the symbol a occurs and the second class will contain all the combinations in which a does not occur. The number of combinations in the second class is $_4C_3$, because by excluding a we have only four symbols left at our disposal. As to the first class, we notice that each combination of this class is obtained by adjoining a to a certain combination of *two* symbols taken among b, c, d, e. It becomes clear that there are $_4C_2$ combinations in the first class. Hence,

$$_5C_3 = {_4C_2} + {_4C_3}.$$

We may check these statements by listing the combinations of the first class:

$$abc \quad abd \quad abe \quad acd \quad ace \quad ade$$

and those of the second class:

$$bcd \quad bce \quad bde \quad cde.$$

Similar reasoning (which the student should repeat with several examples) shows that, for instance,

$$_7C_4 = {_6C_3} + {_6C_4}$$
$$_8C_3 = {_7C_2} + {_7C_3}.$$

To get the general relationship we use the same procedure except that instead of letters it is now more convenient to use the numbers from 1 to n and find the number of combinations of m numbers we can select from these n numbers. We divide the combinations into two classes: the first comprising all those which contain the number 1 and the second all which do not contain 1. To accompany 1 in the first class we have only $(n-1)$ numbers to choose from and $(m-1)$ to choose for any combination. Hence the number of combinations in the first class is $_{n-1}C_{m-1}$. In the second class we have $(n-1)$ to choose from and m to choose for any combination. Hence the number of combinations in the second class is $_{n-1}C_m$ and we have

(1) $$_nC_m = {_{n-1}C_{m-1}} + {_{n-1}C_m}.$$

Then, to find $_5C_3$ we apply this formula to $_4C_2$ and have

$$_4C_2 = {_3C_1} + {_3C_2}$$

which shows that

$$_5C_3 = {_4C_2} + {_4C_3} = {_3C_1} + {_3C_2} + {_4C_3}.$$

To reduce this further, notice that we have proved (problem 2) that $_5C_2 = {_5C_3}$ and this same argument gives $_7C_2 = {_7C_5}$ and $_8C_3 = {_8C_5}$, for example. In general,

(2) $$_nC_m = {_nC_{n-m}}.$$

Apply this general result to $_3C_2$ and $_4C_3$ and have

$$_5C_3 = {_3C_1} + {_3C_1} + {_4C_1} = 3 + 3 + 4 = 10.$$

This shows how the two general properties (1) and (2) of the symbol $_nC_m$ together with the fact that $_nC_1 = n$ serve the purpose of evaluating the symbol $_nC_m$. But even this process would make the computation of $_{100}C_{81}$ quite long. We proceed to further investigation. We form the following table:

$$
\begin{array}{cccccc}
& & 1 & & & \\
& 1 & & _1C_1 & & \\
& 1 & _2C_1 & & _2C_2 & \\
1 & _3C_1 & & _3C_2 & & _3C_3 \\
1 & _4C_1 & _4C_2 & & _4C_3 & _4C_4 \\
1 & _5C_1 & _5C_2 & _5C_3 & _5C_4 & _5C_5 \\
\end{array}
$$

. . . .

Noticing that $_1C_1 = {}_2C_2 = {}_3C_3 = \cdots = 1$ we can write our table in the form

$$
\begin{array}{ccccccc}
& & & 1 & & & \\
& & 1 & & 1 & & \\
& 1 & & _2C_1 & & 1 & \\
1 & & _3C_1 & & _3C_2 & & 1 \\
1 & _4C_1 & & _4C_2 & & _4C_3 & 1 \\
1 & _5C_1 & _5C_2 & _5C_3 & _5C_4 & & 1 \\
\end{array}
$$

. . . .

Relationship 1 tells us, for instance, that

$$_3C_2 = {}_2C_1 + {}_2C_2 = {}_2C_1 + 1$$
$$_4C_2 = {}_3C_1 + {}_3C_2 \quad \text{and} \quad _5C_3 = {}_4C_2 + {}_4C_3.$$

Notice that in each of these cases each C is the sum of its nearest neighbors in the line above. In fact, the nearest neighbors of $_nC_m$ in the line above it are $_{n-1}C_{m-1}$ and $_{n-1}C_m$ and we know from formula (1) that $_nC_m$ is the sum of these. This, you may recall, was exactly the way in which we obtained Pascal's triangle. Hence the above triangular array of C's can be matched exactly with the Pascal triangle. For instance, $_7C_3$ is in the same position as 35 in the Pascal triangle and hence is equal to it.

From these results we shall derive the following formula

(3)
$$_nC_m = \frac{n!}{m!(n-m)!}$$

where $n!$ means $1 \cdot 2 \cdot \cdots \cdot n$ and $n \geq m > 0$. The symbol $n!$ is called **n factorial**.

There is one thing about this formula that must be settled in the beginning. If it happens that $m = n$, our formula becomes

$$_nC_n = \frac{n!}{n!0!}.$$

We see that $_nC_n$ has the value 1 for it is the number of combinations of n things taken n at a time. Hence, in order to make our formula fit this case we must define $0!$ to be 1. This is not in conflict with our definition above since the definition does not make sense if n is zero. Though this definition of $0!$ could be used to give a value of $_nC_0$ such a value is not especially useful; hence we omit it.

In order to prove formula (3) it is first necessary to see that it holds for small values of n and m. For instance, for $n = 4$, the values of $_4C_m$ for $m = 1, 2, 3, 4$ are 4, 6, 4, 1 which are the last four entries in the fifth line of the Pascal triangle. The reader should test the formula for $n = 1, 2, 3$. In a similar fashion, suppose one checked the formula for $n = 1, 2, 3, \cdots, k - 1$, that is, for the first k rows of the Pascal triangle. We now show by computation that it must hold for the next row. (At this point, any timid reader may put $k = 10$ here and throughout the rest of the proof, before following it through with k.) Notice first, that in formula (3) the factorial in the numerator is the first subscript of $_nC_m$ and those in the denominator are the second subscript and the first minus the second. We thus know that

$$_{k-1}C_m = \frac{(k - 1)!}{m!(k - 1 - m)!}$$

and

$$_{k-1}C_{m-1} = \frac{(k - 1)!}{(m - 1)!(k - m)!}$$

provided $m - 1 > 0$ and $m - 1 \leq k - 1$, that is,[1] $m > 1$

---

[1] The case $m = 1$ causes no trouble, for formula (3) holds for $m = 1$ no matter what n is since $_nC_1 = n$ and $n!/(n - 1)! = n$.

and $m \le k$. Then from formula (1) with n replaced by k we have

$$_kC_m = \frac{(k-1)!}{(m-1)!(k-m)!} + \frac{(k-1)!}{m!(k-m-1)!}.$$

Now notice that

(4) $m! = (m-1)!m$ and $(k-m)! = (k-m-1)!(k-m)$.

Thus

$$_kC_m = \frac{(k-1)!}{(m-1)!(k-m-1)!(k-m)} + \frac{(k-1)!}{(m-1)!(k-m-1)!m}$$

$$= \frac{(k-1)!}{(m-1)!(k-m-1)!}\left\{\frac{1}{k-m} + \frac{1}{m}\right\}$$

$$= \frac{(k-1)!}{(m-1)!(k-m-1)!} \frac{m+k-m}{m(k-m)}$$

$$= \frac{(k-1)!k}{[(m-1)!m][(k-m-1)!(k-m)]}$$

which, by formula (4), yields

$$_kC_m = \frac{k!}{m!(k-m)!}$$

which is exactly formula (3) with n replaced by k. Even though it may have seemed slightly more difficult to use k instead of 10 or some other number in our proof, it has the distinct advantage that our result holds for whatever positive integer k is. This shows that if formula (3) holds for $n = 10$, it must hold for $n = 11$; then since it holds for $n = 11$, it holds for $n = 12$ and so on. Since we know it holds for the first values of n, it must now hold for all values.

EXERCISES

1. By writing down all the different combinations find the values of $_6C_4$ and $_6C_2$. Can you see any way in which one could have been found from the other?

2. Use the Pascal triangle to find the values of $_8C_4$, $_6C_3$, $_{10}C_2$, and $_8C_5$.

3. What property of the Pascal triangle corresponds to the fact that $_nC_m = {}_nC_{n-m}$?

4. Assume formula (3) for $n = 30$ and use formula (1) to prove formula (3) for $n = 31$.

5. Find the number of different 5-card hands one can draw from a complete deck of 52 cards.   *Ans.* 2,598,960.

6. How many different bridge hands can be dealt? Estimate the magnitude of your answer.   *Ans.* About 635,000,000,000.

7. Suppose we have the rectangle of Fig. 6:2 whose dimensions are 3 units by 4 units, the vertical and horizontal units being of different

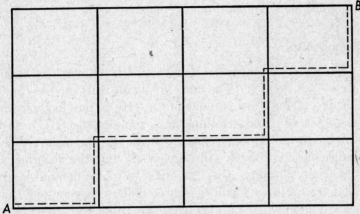

Fig. 6:2

lengths. In how many different ways can one move from A to B if one is allowed to move either one unit to the right or one unit up each time?

SOLUTION: It is not hard to see first of all that there will be $3 + 4 = 7$ moves. Let us denote them by 1, 2, 3, 4, 5, 6, 7. The way from A to B is completely determined by choosing the vertical moves. For instance, (2, 5, 7) denotes the way of the dotted line on the diagram because the second, fifth, and seventh steps are vertical. [Show on the diagram the way (3, 6, 7).] Now (2, 5, 7) is a combination of three symbols out of seven and since every such combination determines one and only one way of reaching B from A and since conversely each way determines one and only one combination of three symbols (out of seven), the total number of ways is $_7C_3 = 35$. The student should prove that the answer is also $_7C_4$ by considering horizontal steps.

8. A checkerboard is so placed that a black square is in the lower left-hand corner. A checker is placed on the third black square counting from the left in the first row and it is to be moved on black squares to the corresponding square in the seventh row. In how many ways can this be done assuming that no other pieces are on the board and that the checker is not a king, that is, it can move only forward? What would your answer be if it were to be moved to the second black square counting from the left in the eighth row? If to the third black square from the left in the eighth row?

* 9. Chess is played on all the squares of a checkerboard. A pawn beyond the second row has at any time at most three possible moves; to the square directly in front of it and, if it can take a piece, to one of the squares immediately to the right or left of the one in front of it. A pawn is placed at the left end of the fourth row. In how many different ways can it reach the eighth row?

3. Probability.

We often hear the statement: "His chances of being elected are small"; "The odds are against his success"; and the like. These are equivalent to "He is not likely to be elected" and "I do not think he is going to succeed." Such statements seem much more specific in the form: "The chances are 1 to 3 he will be elected" but the accuracy is somewhat spurious. One can make a really definite statement by saying, "I will bet three dollars to your one that he will lose." This shows that we think our chances of guessing correctly are three to one, but of course one cannot in such a situation determine exactly what his chances are.

There are, however, cases in which one has a definite, well-based reason for a statement about chances. If I threw one die I should say that my chances of throwing a 1 would be one to five. My reason for this statement would be that there are six faces of a die of which only one is a 1. If the die were honest I should expect throwing a 1 to be "equally likely" to throwing any one of the other five. Hence throwing a 1 should be one-fifth as likely as throwing something else. "I feel it in my bones that in the long run" a 1 should appear one-sixth of the time and might support this statement by saying that if it did not, I should suspect that the die was loaded. This circuitous "reasoning" leads us nowhere. The mathematician avoids this tangle by saying that if a thing can happen in any one of n equally likely ways and if, of these, f are favorable, then the **probability** of its happening favorably is

$$f/n.$$

Hence if in throwing a die one face is as likely to appear as any other, the probability of throwing a 1 is 1/6. Notice that the odds in this case are one to five. Similarly odds of one to ten are equivalent to the probability of 1/11.

Even though we state without qualification that the probability is 1/6 or 1/11 or what-not, we must not lose sight of the fact that we do so on the assumption that one thing is **equally likely** to another. This appears in the determination of the probability of throwing two heads with a pair of coins. We might reason: one of three things may happen: I may throw two heads, two tails, or one head and one tail; hence the probability of throwing two heads is 1/3. But this is wrong because we are twice as likely to throw one head and one tail as may be seen by noticing that one head and one tail may appear in one of two different ways each of which (we assume) is equally likely to throwing two heads. Hence we find that the probability of throwing two heads with a pair of coins is 1/4.

Such probability which we determine by calculation we call **a priori probability** as opposed to **a posteriori probability** which is determined as a result of experiment or statistical study. We shall consider the latter in the next section.

There is one common misconception of probability that is very hard to scotch. It is the notion that the "law of averages" is Nature's guardian who keeps a stern eye on her to see to it that whenever at night she misbehaves in one direction, she makes up for it the next morning by misbehaving in another direction. For example, suppose you have tossed a coin nine times and every time it has turned up heads. What is your chance of throwing tails the next time? From one point of view the chance is very small, for since the probability that an "honest" coin fall heads nine times out of nine is 1/512, I should be led to suspect that the coin is not "honest" and is more likely to fall heads than tails. But if the coin is honest, that is, if any time it is as likely to fall heads as tails, then one of the times is after it has fallen heads nine times and the probability of its falling tails the tenth time is one-half.

Example 1. Find the probability of throwing a 7 with a pair of dice; an 8; a 12.

SOLUTION: One can get a total of 7 in six different ways as is shown in the following table:

Die I	1	2	3	4	5	6
Die II	6	5	4	3	2	1

One can get a total of 8 in five different ways, which the reader should list, and a 12 in just one way. To find the total number of ways a pair of dice can fall, we could list them all and perhaps the reader will prefer to do it that way. But we can save ourselves labor by seeming to change the subject and talking about three towns A, B, and C. Number six roads from A to B and six from B to C. Being undecided which road to take, we toss the dice and let the first die tell us what road to take from A to B and the second what road from B to C. Then, clearly, the number of different routes from A to C by way of B will be the number of different ways the dice can fall. But the number of routes is $6 \cdot 6 = 36$. Hence, on the assumption that each way in which the dice may fall is equally likely to occur, the probability of throwing a 7 is $6/36 = 1/6$, of an 8 is $5/36$, and of a 12 is $1/36$.

Example 2. What is the probability that if four dice are tossed at least two of them will show a 6?

SOLUTION: Notice first of all that "at least" means that two 6's shall appear. In other words, we wish the probability that exactly two 6's, exactly 3, or exactly 4 will appear. To do this on the usual assumption that one appearance is as likely as any other, we need to compute the number of ways exactly two 6's may appear, add to this the number of ways exactly three 6's may appear and four 6's may appear.

First we compute the number of ways exactly two 6's

may appear. Again we change the subject and think of five towns A, B, C, D, and E and six roads from each to the next. Here we must take exactly two roads numbered 6 in our journey from A to E. If we choose roads 6 from A to B and from B to C, then the roads numbered 6 from C to D and D to E are to be avoided and we have only five roads open for each of the last two legs of the journey. Hence there are $5 \cdot 5$ or 25 permissible routes from C to E by way of D. Thus, taking roads numbered 6 only from A to B and from B to C we have 25 routes. The story would be the same if we had elected to take roads 6 from A to B and from D to E, and so forth. Hence the total number of routes using exactly two roads 6 is 25 multiplied by the number of ways we can pick two from four routes numbered 6. Our answer for this part, then, is

$$_4C_2 \cdot 25 = 150.$$

It is easier to compute the number of ways exactly three 6's appear for if we choose road 6 from A to B, B to C, C to D we have five roads not numbered 6 to choose from for the journey from D to E. There are four different towns we can leave on a road not numbered 6 and our answer for this part is $4 \cdot 5 = 20$.

There is just one journey using all routes 6 and hence our total permissible routes are in number

$$150 + 20 + 1 = 171.$$

The total number of routes is $6^4 = 1296$. Hence the probability of throwing at least two 6's in a toss of four dice is

$$171/1296.$$

Example 3. Find the probability that in tossing ten coins exactly three are heads.

SOLUTION: If the first three are heads, the last seven must be tails if the throw is what we wish it to be and there is only one way in which the seven can be tails. Hence the number of ways one can have three heads is the same as the number of ways we can pick three out of ten to be heads, that is $_{10}C_3 = 120$. On the other hand, there are $2^{10} = 1024$ ways in which the ten coins can fall, for we may think of eleven towns in a row each connected with its predecessor by two roads. There will be two routes connecting town 10 with town 11, 2^2 routes from town 9 to 11, 2^3 from 8 to 11, $\cdots$, 2^{10} routes from town 1 to town 11. Hence the probability of throwing exactly three heads out of the ten is

$$_{10}C_3/1024 = 120/1024 = 15/128.$$

Example 4. Find the probability of drawing from a deck of playing cards a five-card hand containing two of one kind (that is, denomination) and three of another; e.g., two kings and three queens. (This is called a **full house** in the game of Poker.)

SOLUTION: Since we know from the result of Exercise 5 of section 2 that there are 2,598,960 five-card hands, this will be the denominator of the value of our probability and the numerator will be the number of different full houses in the deck. We first find the number of three-card hands containing three cards of one denomination. Since there are 13 different denominations and 4 such different three-card hands for each denomination, there will in all be $4 \cdot 13 = 52$ different three-card hands whose cards are all three of the same denomination. The denomination of the remaining two cards may be any one of the remaining twelve, and for each denomination there will be $_4C_2$ or 6 different pairs. Hence there are $12 \cdot 6 = 72$ possible pairs after the three of a kind have been selected; in other words, for each three of a kind there are 72 possible pairs to complete the full house. Thus the number of full houses is $52 \cdot 72 = 3744$ and the probability of drawing such a hand is $3744/2,598,960$ which is approximately $3/2000$.

Example 5. A man playing Draw Poker holds the following hand: the king of hearts and of diamonds, the queen of hearts and of diamonds, and the jack of hearts. He is undecided whether to discard the jack and hope to draw a queen or king, thereby having a full house, or to discard the two diamonds and hope to get a "straight" (five cards in sequence) or a "flush" (all cards of the same suit). Which is more likely: that discarding the jack he get a full house, or discarding the diamonds one of his other hopes is realized? Assume he has the rest of the deck to draw from.

SOLUTION: For the first alternative, there are two queens and two kings remaining in the deck and hence his probability of getting a full house is 4/47. For the second, notice that to get a straight he must draw a 10 and either an ace or a 9. Hence there are $4 \cdot 8 = 32$ different ways he can complete his straight. If he is to get a flush he must draw two hearts out of the ten hearts remaining in the deck. There are $_{10}C_2 = 45$ different such pairs. Now we might carelessly say that there are $32 + 45 = 77$ different pairs of cards which the player can draw, each one of which will yield a straight or a flush. But we have included in both the 32 and the 45 the cases in which the 10 and ace or 9 are hearts, that is, in which the resulting hand is both a straight and a flush, a so-called **straight flush**. These two hands we have counted twice. Hence $77 - 2 = 75$ is the correct number of desirable hands. There are $_{47}C_2 = 47 \cdot 23$ different possible two cards which can be drawn from 47 cards. Hence the probability of getting either a flush or a straight when two cards are discarded is

$$\frac{75}{47 \cdot 23} = \frac{1}{47}\frac{75}{23}.$$

This is less than 4/47. Hence the player should try for a full house. Notice that this conclusion is reached without taking into account another consideration in its favor, namely, that even if neither a queen nor a king are drawn under the first alternative, the player still has two pairs in his hand.

Example 6. Suppose that four persons each toss independently a coin. What is the probability that exactly two of them will fall heads?

SOLUTION: It is easy to enumerate all the possible results as follows:

Person I	h h h h h h h h t t t t t t t t
Person II	h h h h t t t t h h h h t t t t
Person III	h h t t h h t t h h t t h h t t
Person IV	h t h t h t h t h t h t h t h t

where h stands for heads and t for tails. The event of getting exactly two heads occurs in six cases (underlined in the diagram). The total number of possible situations is $16(= 2^4)$, hence the probability

$$6/16 = 3/8.$$

Example 7. What is the probability that if 5 coins are tossed, exactly 3 will fall heads?

SOLUTION: We could again construct a table of possibilities but we prefer to use our knowledge of combinations. Denoting the coins by I, II, III, IV, and V, we mean by the symbol (II, III, V) the fact that the second, third, and fifth coins will fall heads and the others fall tails. There are obviously $_5C_3 = 10$ such symbols and hence there are 10 different occurrences of the event of getting exactly three heads. Thus the total number of possibilities is

$$1 + {_5C_1} + {_5C_2} + {_5C_3} + {_5C_4} + {_5C_5} = 32 = 2^5$$

because we get either no heads (1 possibility) or 1 head ($_5C_1$ possibilities) or 2 heads ($_5C_2$ possibilities) or 3 heads ($_5C_3$ possibilities) or 4 heads ($_5C_4$ possibilities) or 5 heads ($_5C_5$ possibilities). The answer then is

$$10/32 = 5/16.$$

Notice that 2^5 could also be arrived at by methods like those used in the solution of Example 3 above.

EXERCISES

(In all of these, the dice, coins, and cards are assumed to be "honest.")

1. Find the probability of getting the totals of 2, 3, 4, 5, 6, 9, 10, 11 with 2 dice.

2. Find the probability of getting a total of 6 by tossing 3 dice.

3. Prove that the probability that exactly 4 out of 8 coins will fall heads is $_8C_4/2^8$.

4. Find the probability that exactly 2 out of 4 coins will fall heads. Should your answer differ from that of Example 6?

5. Generalize the above results for the case of m heads out of n coins.

6. Prove that the probability that at least 2 out of 4 coins will fall heads is

$$\frac{_4C_2 + {}_4C_3 + {}_4C_4}{2^4} = \frac{11}{16}.$$

7. Find the probability that at least 3 out of 6 coins will fall heads.

8. Which is more probable, getting a total of 7 with 2 dice or getting at least 4 heads by tossing 5 coins?

9. Without using the results of Example 2 solved above, find the probability that if 4 dice are tossed, at most one will show a 6. Why does this give another method of answering the question of Example 2?

10. Find the probability that at least 3 out of 6 dice will be 5's.

11. Find the probability that *exactly* 2 out of 5 dice will be 6's.

12. What is the probability that, tossing a coin and 2 dice, one will get a head and a total of 8 on the dice?

13. Find the probability that, tossing 4 coins and 4 dice, at least two heads and exactly one 6 will occur.

14. Find the probability that in tossing 10 coins exactly 4 are heads.

15. Little Lulu tore all the labels from 15 cans her mother brought home from the store and thoroughly mixed them. The cans were all of the same size; 4 contained corn, 6 beans, and 5 tomatoes. Her mother picked out a can and opened it regardless. What is the probability (1) that it contained corn, (2) that it contained beans or tomatoes?

16. Find the probability that from a deck of playing cards one will draw a 13-card hand

    a. In which all the cards are of the same suit,

    b. Containing four aces,

    c. Containing the A, K, Q, J, 10 of one suit.

17. Find the probability that from a deck of playing cards one draws a 5-card hand

    a. Containing 4 of a kind, that is, 4 of the same denomination,

    b. All of whose cards are of the same suit: a flush,

    c. Whose cards are in sequence: a straight. An ace may be used at the top or bottom of the sequence.

18. A true-false test contains 10 questions. What is the probability that marking the answers at random one will have 60% or more correct? Would the answer be the same if there were only 5 questions?

19. Answer the question of Example 5 solved above for the hand containing the queen of hearts and of diamonds, the jack of hearts and of diamonds, the ten of hearts.

4. A posteriori probability.

Using the methods of Example 3 and Exercise 14 we could complete the following table of the probability of throwing 0 heads, 1 head, 2 heads, $\cdots$, 9 heads, 10 heads in one throw of 10 coins.

Table A

$$\frac{1}{1024}, \frac{10}{1024}, \frac{{}_{10}C_2}{1024}, \cdots, \frac{{}_{10}C_2}{1024}, \frac{10}{1024}, \frac{1}{1024}.$$

The values of the numerators are the numbers of the last line of the Pascal triangle given in section 6 of Chapter IV.

Table B

Number of heads	0	1	2	3	4	5	6	7	8	9	10
Frequency	1	10	45	120	210	252	210	120	45	10	1

Such a table as this might also be the result of such an experiment as the following. Suppose 32 persons toss a set of ten coins 32 times each and record the number of times the set of ten showed no heads, one head, etc. If they then pooled their results we should expect not a very great deviation from Table B since the set of ten coins would have been tossed $32 \cdot 32 = 1024$ times. In fact, for any large number of tosses we should expect the ratios of the various frequencies to the total number tossed to be not far from the ratios of Table A. Notice that Table 23 in Chapter V is in accord with this statement, for there the results were so close to the "theoretical" values of Table A above which can be obtained without experiment (a priori probability) that it would not much matter whether we based any predictions on the experimental or theoretical table. But the points of view are somewhat different. In the case of the theoretical table we calculated in how many ways each number of heads could occur and, assuming vaguely that each way was "just as likely to happen" as any other way, divided the number of ways in which the specified number of heads could occur by the total number of ways in which the coins could fall. On the other hand, the experimental table showed us how the coins fell in a particular experiment. Then our computation of the probabilities is based on the supposition that, "in the main," things will continue to happen as they have happened in the past. Due to the difficulty of finding a priori probability in most situations, statements of probabilities are more apt to be based on experimental or observational tables. And these considerations, vague as they are, are often sufficiently accurate for useful predictions. The mortality table, for example, is exact enough to provide the foundation for life insurance. Though the insurance company does not know which of a group of 100,000 individuals aged ten will die between his thirtieth and fortieth birthdays, it does know that about 7300 are practically certain to die within that period of years — barring a surprising advance in medicine or a disastrous epidemic or war before or during that period. To say that the probability that a boy aged ten will die between his thirtieth and fortieth years is 7300/100,000 is merely expressing in terms of the individual the fact that 7300 out of 100,000 such people die. This statement is correct only if we think of it as another means of expressing what the mortality table tells us about 100,000 people. We could apply it to an individual in the sense of the previous section only if one individual were as likely to die as another. Lacking that knowledge, we cannot say truthfully that any individual aged ten which you may select has 73 out of 1000 chances of dying between thirty and forty.

This point may be clarified by another example. A man is running a dart booth at a fair and finds from experience that one-tenth of those who try succeed in hitting a balloon with a dart. He could then say quite properly that the probability of anyone winning a prize is 1/10 and he can use this as a basis for the amount he will charge and the value of his prizes. But his statement does not in the least mean that whoever steps up to the board has one chance in ten of securing a prize because not all are equally skillful. Some may win whenever they play and others never; the probabilities are respectively 1 and 0.

The measure of both kinds of probability is a number not less than 0 and not greater than 1 associated with each of the recognized ways in which an event may turn out. In terms of the graph, this number is the ratio of the area of the appropriate rectangle in the histogram to the sum of the areas of all the rectangles.

EXERCISES

1. What is the probability that
   a. A person aged 10 will be alive at age 70?
   b. A person aged 65 will be alive at age 75?
   c. A person aged 40 will be alive at age 80?
2. What is the probability, in tossing 10 coins, of getting
   a. Exactly 7 heads?
   b. Exactly 5 tails?
   c. Not less than 8 heads?
   d. Either 2 or 3 heads?
3. Use Table 21 in Chapter V to find the probability that a call will last
   a. Just 450 seconds;
   b. Not more than 150 seconds;
   c. Not less than 750 seconds;
   d. More than 750 seconds.
4. Use either the mortality table or the table computed in Exercise 3, section 8 of Chapter V, to find the probability that a person aged:
   a. 10 will die between his 30th and 35th birthdays;
   b. 30 will die before he is 35;
   c. 60 will not die before he is 70;
   d. 25 will die between his 60th and 70th birthdays.
5. Do you agree with the conclusions in the following paragraph which appeared in a popular magazine under the heading: "One Third of Your Children Doomed"? It reads: "Being of a cheerful turn of mind, we have tried to leave to insurance companies the somber task of charting life's overwhelming hazards. One particular hazard, however, has increased so steadily that we can hardly be held a spoilsport for calling the attention of 1935's parents to the fact that their children, according to the latest estimates, stand one chance in three of meeting death or serious injury from that deathlike convenience, the motor car. Take a room containing three children: one of them is destined to be killed or badly hurt by a car before he has completed his normal life span. Statistically,

the motor car is life's ugliest joke; its toll makes war seem like a spring outing."

6. Is the above quotation an argument for having no more than two children?

7. The story is told that farmer Silas figured that since a recent flood a certain rickety bridge near his farm has one chance in ten of collapsing when he was on it. One day, recalling that since the flood he had crossed the bridge nine times, he realized that he had used up all his chances and hence would in the future keep off the bridge. Assuming that the probability figure was correct, would you behave as he did after crossing the bridge nine times? Why?

5. Mathematical expectation.

It is rather startling to notice that those who make most use of probability are of two classes diametrically opposite in their attitude toward money: the reckless gambler who will stake his all on the turn of a wheel, and the timid conservative who provides for every possible mishap by taking out all available insurance; in the first case all losers pay the winner and in the second all winners pay the loser.

The use which they make of probability is by way of what is called **mathematical expectation**. We illustrate it first by the description of a simple game. Suppose seven people choose one of their number as the banker. Each of the remaining six selects one of the numbers 1, 2, $\cdots$, 6 as his number, no two having the same number. Then the banker throws one die and the number which turns up determines the winner. The question is, if the winner is to receive $6, how much should each pay the banker for the privilege of playing? The obvious and correct answer is $1 since then the banker will receive just enough money to pay the winner. (As a matter of fact they could dispense with the banker by forming a pool.) If the banker were running the game he might charge each player $1.25 and make a tidy profit. Notice that the $1 which a player should pay if the banker breaks even is the product of the probability of winning and the amount of the prize. The **mathematical expectation** is defined to be the product of the amount to be won and the probability of winning it in problems like this when there is only one way of winning and only one amount to be won. Further modifications of this are indicated in examples below. In this particular case the mathematical expectation is $1 and at that figure each player should expect neither to lose nor gain much after many games of an evening. On a payment of $1.25 he should expect to lose 25 cents per game for the evening as a whole. As a matter of fact, the overcharge of 25 cents is small compared to the usual practices of those who run gambling games.

In the realm of insurance essentially the same situation would arise. If a fire-insurance company found that over a period of time one out of every sixty cars in a certain community burned, it would expect to charge a premium of 1/60th of the amount of the policy plus expenses and should then in the long run take in just enough money to pay all claims plus expenses, on the assumption that fires *will do* the same damage that they *have done*. Here the insurance company assumes the risk and, quite rightly, charges for it. On the other hand, companies which operate fleets of trucks usually assume the risks themselves.

Example 1. A man draws one card from a deck of playing cards. If he draws a red ace he is to receive $2, if a king $1, otherwise nothing. How much should he pay to play?

SOLUTION: To clarify the situation we at first change the problem slightly and suppose that two persons, A and B, play, the former winning only with a red ace and the latter only with a king. The probability that A will win is then $2/52$ and his expectation is $2 \cdot 2/52 = 1/13$ of a dollar. Similarly the probability that B will win is $1 \cdot 4/52 = 1/13$ of a dollar. Together they would pay $2/13$ dollars. Now in the problem as given, the single player wins when either of the players A or B would win and hence he should pay what they together pay. Therefore, he should pay $2/13$ dollars which is $.15 to the nearest cent. This is called his **mathematical expectation.**

Example 2. A man 40 years old wishes term insurance to the amount of $1000 for five years. That is, if he dies within 5 years the company pays $1000 to his heirs; if he lives for that length of time the company pays nothing. He chooses to pay for his policy in a single premium at the beginning of the 5-year period. What should he pay, neglecting any charge the company may make for its expenses?

SOLUTION: Table 22 in Chapter V shows that out of 78,106 persons alive at 40 years of age, 74,173 are alive at age 45. Hence his probability of dying is 3933/78,106 = .050353. Thus his premium should be 1000(.050353) or $50.35.

NOTE: Term insurance is not the most common kind of insurance. If, at the beginning of each year, one procured term insurance for that year, he would pay each year a higher premium than the last since his probability of dying increases (this is not necessarily so in any individual case but it is true statistically for a large number of cases). In ordinary life insurance, the payments are the same one year as the next and in one's earlier years he accumulates a reserve to offset the increase in cost of insurance as he grows older. On the other hand, the usual group insurance is term insurance in which (unless the employer contributes) the younger men pay more than they would pay for individual term insurance and the older men pay less.

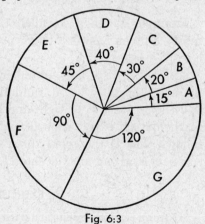

Fig. 6:3

Example 3. An arrow is spun on a cardboard disc divided into seven sectors as in the figure. Players put money on any or all of the sectors. If the arrow stops on a dividing line, no one loses and it is spun again. If it stops in sector A, each person having money in A is given a times as much money as he put down in that sector and the banker collects the money in all the other sectors. If the arrow stops in B, b times as much is paid to all whose money is in B, and so forth. Assuming the arrow is just as likely to stop in one spot as in another, determine the correct values of a, b, c, d, e, f, and g.

SOLUTION: Sector A is $15/360 = 1/24$ of the total area. Hence anyone putting money in section A has a mathematical expectation of $1/24$th of what he gets if he wins. Hence if he wins he should get 24 times what he put in. Similarly, the values for b, c, d, e, f, and g are 18, 12, 9, 8, 4, 3. We may check our results by finding that 1 is the value of the sum

$$1/24 + 1/18 + 1/12 + 1/9 + 1/8 + 1/4 + 1/3.$$

Notice that those who put money in A, for instance, actually *gain* 23 times as much as they put in, for each dollar placed in the sector is returned accompanied by 23 other dollars. Thus the *odds* for sector A would be 1:23.

Example 4. In a so-called true-false test the examinee marks a given answer "true" or "false." How should such a test be scored?

SOLUTION: If a large number of persons took the test, marking the answers at random, we should expect about half the answers to be right and half to be wrong. Since the mathematical expectation of each person should be zero, it follows that if the score for a correct answer is $+1$, the score for a wrong answer should be -1.

EXERCISES

1. A bag contains 8 white and 2 red balls. A man draws from the bag and is to receive $1 for each red ball he draws but nothing for a white ball. What should he pay to play if he

  a. Draws once?

    b. Draws twice, not replacing the first ball drawn?

    c. Draws twice, replacing the first ball drawn?

    2. A hat contains 10 discs numbered from 1 to 10. A player drawing receives the number of dollars corresponding to the number on the disc. How much should one pay to draw once from the bag?

    3. How much should a man aged 50 pay in a single premium for a 5-year $1000 term-insurance policy? Make no allowance for the expenses of the insurance company.

    4. In a certain city one out of every 5000 cars is stolen and not recovered. Allowing no expenses for the company, what should be the premium for a $1000 policy?

    5. A board has squares numbered 2 to 12. Players put their money on one or several squares. Two dice are tossed and the sum of the spots showing determines the winning square. What odds should be marked on each square? (See Example 3 worked out above.)

    6. A game, one of whose names is "chuck-a-luck," is played as follows. A player chooses a number from 1 to 6 inclusive and throws 3 dice. If his number appears on just 1 die he receives 10 cents, if on just 2 he gets 20 cents, if on all 3, 30 cents. Otherwise he is paid nothing. How much should he pay to play?

    7. Suppose the game described in the previous exercise were altered so that if the player's number appeared just once he had his money returned and 5 cents in addition, if it appeared exactly twice he received 10 cents beyond what he paid, and if all three times, 15 cents. Otherwise he is to lose what he paid. How much should he pay to play?

    8. The statistics of an insurance company show that over a period of years one out of every hundred cars in a certain community burns. However, it has found that in section A in the town twice as many cars burn in any year as in the remainder of the town, even though there are just as many cars in section A as outside it. Allowing for no company expenses, what should be the fire-insurance rates in the two parts of the town?

    9. If a true-false test of 10 questions is scored as in Example 4, what is the probability that a person marking the answers at random should get 0 for a total score?

    10. If a "multiple-choice test" has 4 answers listed for each question, exactly 1 of which is correct, how should it be scored?

6. Topics for further study.

For this subject the topics are so varied and interlaced that it will be better for you to select them yourself. See reference **25**, Chap. 7, and reference **19**, Chap. 14.

CHAPTER VII

Mirror Geometry

or the
Theory of Enantiomorphous Figures

JABBERWOCKY

'Twas brillig, and the slithy toves
Did gyre and gimble in the wabe;
All mimsy were the borogoves,
And the mome raths outgrabe.

"It seems very pretty," she said when she had
finished it; "but it's *rather* hard to understand!"
(You see she didn't like to confess, even to herself,
that she couldn't make it out at all.) "Somehow it
seems to fill my head with ideas — only I don't exactly know what they are!",
— From *Through the Looking Glass.*

1. Introduction.

The preceding chapters have dealt chiefly with numbers or
with the study of simple geometric figures by means of numbers. We now turn to some topics in geometry where
numerical relationships are either absent or occur only
incidentally.

In the algebraic work you were expected to remember only
a few of the more elementary properties of numbers and
algebraic expressions. Here we ask you to remember from
your study of Euclidean geometry only some of the simpler
theorems. Thus we shall want to know that it is possible to
draw one and only one line perpendicular to a given line and
through a given point. A theorem we require is that two
triangles are congruent if two sides and the included angle
of one are equal to the corresponding sides and the included
angle of the other. A few other results from Euclidean
geometry will be used in this chapter.

2. Symmetry and reflection.

A vase shaped as in Fig. 7:1 is generally regarded as being more beautiful than one shaped as in Fig. 7:2. In many sorts of beautiful objects there is a quality we somewhat loosely

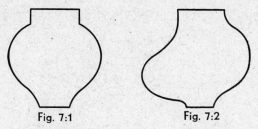

Fig. 7:1   Fig. 7:2

refer to as "symmetry." Even in a painting where the two halves of the picture certainly do not "match" there are often certain elements of "symmetry." In architecture, particularly that derived from Roman models, the symmetry of the façade is often complete.[1]

The vague notion of symmetry which we have just suggested is, of course, of no use to the mathematician until he has distilled its essence and preserved his analysis of its structure in a definition. Accordingly we first simplify our discussion by limiting it to figures in a plane and then proceed to set up our definition.

One significant fact must not escape our attention if we are to think clearly about symmetry. Symmetry (of the kind we are here discussing) is *with respect to some particular straight line*. Thus our first glance at Fig. 7:3 may not reveal the fact that it is symmetric but a second inspection, perhaps aided by turning the page partly around so that the line AB is vertical, shows it to be symmetric. It is furthermore evident that the part of the figure on one side of AB may be obtained from the part on the other side. We say that this is accomplished by **reflection** in the line AB or that one half of the figure is the **mirror image** of the other half.

Two figures which are obtained from one another by reflection in a line are sometimes called an **enantiomorphous**

[1] See topic 2, at the close of Chapter VIII.

pair. The same term may be applied to physical objects such as a pair of gloves or a pair of shoes. When we speak of reflecting a solid object, we are, of course, using a plane rather than a line to define the image.

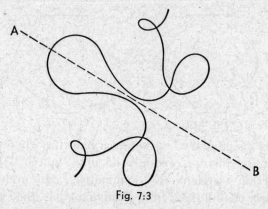

Fig. 7:3

It often happens in mathematics that an attempt to understand fully some vaguely comprehended idea results in the discovery of some more fundamental concept. So it is here. Only certain figures have the property of being symmetric in a line, but *any* figure may be reflected in any given line. Symmetry is a descriptive or static property which a figure may or may not have, while reflection in a line is an operation or *transformation* which we may apply to any figure we like.

It is rather useful to think of a geometric transformation as a rule which enables us to establish corresponding figures in much the same way as the formulas of Chapter V enabled us to construct tables of corresponding numbers. Thus the accompanying table describes the relationships shown in Fig. 7:4.

Original figure	Point P	Triangle ABC	Circle D	Arc EF	Point Q
Figure obtained by reflection in l	Point P'	Triangle $A'B'C'$	Circle D'	Arc $E'F'$	Point Q

As yet we have never really said just how the image of a given figure is to be obtained. Evidently it will be enough to tell how to construct the image of any point, since this will determine the image of any figure made up of points.

Construction: To obtain the image in a given line l of a given point P not on l, draw a line through P perpendicular to l, as in Fig. 7:4. Call the foot of this perpendicular F. Extend the line segment PF to a point P' where $PF = FP'$. Then P' is said to be the *image* of P.

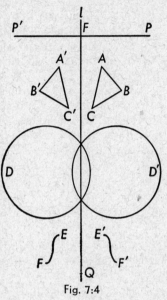

In this construction we explicitly assumed that P was not on l. (Why was this necessary?) In case P *is* on l we define its image to be itself.

With our description of the operation of reflecting figures in a straight line now complete, we can define the property of being symmetric in a line.

Definition: A figure is said to be symmetric in a line l if it is identical with its image in l.

Fig. 7:4

EXERCISES

1. Name some commonly occurring pairs of enantiomorphous objects.

2. In what line or lines are the following figures symmetric?
   - a. An isosceles triangle.
   - b. An equilateral triangle.
   - c. A circle.
   - d. A rectangle.
   - e. A square.

3. When we look in a mirror we see that in the image right and left are interchanged. Why are not up and down interchanged too?

4. Show that the image in l of the image in l of a point P is P itself. What property of the table for Fig. 7:4 does this result imply? Can you find numbers a and b which are such that the table

x	2	3	?
$x' = ax + b$	$2a + b$	$3a + b$	3

will have this same property? (The line l is the line $x = 0$.)

5. What effects on the x and y coordinates of a point are produced by reflecting the point in the x-axis? In the y-axis?

6. Determine the image in a line *l* of each of the following:
 a. A straight line parallel to *l*.
 b. A straight line perpendicular to *l*.
 c. A straight line neither parallel nor perpendicular to *l*.
 d. The line *l*.
 e. A line segment of which *l* is the perpendicular bisector.
 f. A circle tangent to *l*.
 g. A circle with a diameter along *l*.
 h. A triangle with one side along *l*.

7. Would the following definition be equivalent to the one given above? Why?

A figure is said to be symmetric in a line *l* if it can be divided into two parts in such a way that one part is the image in *l* of the other part.

8. Show that the distance between any two points in the plane is the same as the distance between the images of the points in any given line.

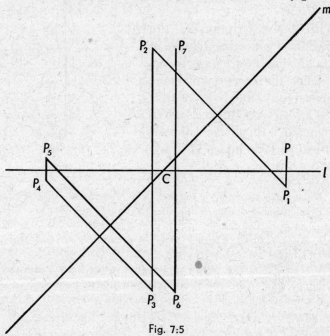

Fig. 7:5

3. Successive reflections.

Let us see what would happen to a point if we were to reflect it successively in two intersecting lines. In Fig. 7:5 we begin with a point *P*, reflect it in a line *l*, obtaining P_1;

then reflect this point in line m (why not in l again?) obtaining P_2; then reflect P_2 in line l, obtaining P_3; and so on. Evidently we might also have begun by reflecting P in m, following this by a reflection in l; and so on as illustrated in Fig. 7:6.

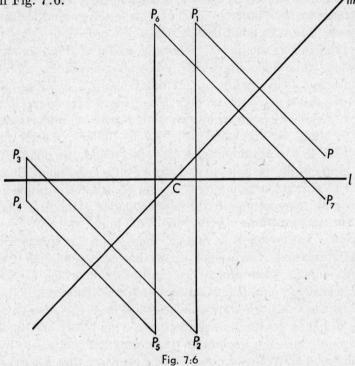

Fig. 7:6

While the points constructed in this way from P "jump about" considerably, even a casual inspection convinces us that they do not wander all over the plane. Indeed, if we notice that the intersection, C, of l and m remains in the same position when reflected in either l or m, we can use the result of Exercise 8 above to conclude that all the points P, P_1, P_2, P_3, etc., are at the same distance from C. This proves the following theorem.

Theorem. Given a pair of lines intersecting in C and a point P different from C, then P and its successive images in the lines all lie on a circle with center at C and passing through P.

If we are to deal readily with the operations of reflection in two lines, we need a convenient way in which to describe the points obtained by successive transformations. For this purpose we shall agree to let the script letter $\mathcal{L}$ stand for the phrase "take the point designated by the preceding symbols and reflect it in the line l." Similarly, we let $\mathcal{M}$ stand for the phrase "take the point designated by the preceding symbols and reflect it in the line m." Thus the point P_1 of Fig. 7:5 is now called $P\mathcal{L}$ and the point P_1 of Fig. 7:6 is now called $P\mathcal{M}$. When we write the expression $P\mathcal{L}\mathcal{M}\mathcal{L}$ we understand it to stand for the point obtained as follows: reflect the point P in l to obtain a point $P\mathcal{L}$; then reflect the point $P\mathcal{L}$ in m to obtain the point $P\mathcal{L}\mathcal{M}$; finally reflect the point $P\mathcal{L}\mathcal{M}$ in l to obtain the point $P\mathcal{L}\mathcal{M}\mathcal{L}$. Our shorthand notation is not only more concise, it gives us concrete symbols $\mathcal{L}$ and $\mathcal{M}$ for the **operations** of reflecting in the lines l and m respectively. The symbols $\mathcal{L}$ and $\mathcal{M}$ then stand for **geometric transformations**. We shall agree that a symbol like $\mathcal{L}\mathcal{M}$ shall stand for the geometric transformation which carries an arbitrary point P into the point $P\mathcal{L}\mathcal{M}$. Thus $\mathcal{L}\mathcal{M}$ is the geometric transformation which results from the successive performance of the transformations $\mathcal{L}$ and $\mathcal{M}$. We shall agree to call $\mathcal{L}\mathcal{L}$ a **geometric transformation** but since (by Exercise 4 above) it has the peculiar property that it carries each point into itself, $(P\mathcal{L}\mathcal{L} = P)$, we call it the **identity transformation** and denote it by $\mathcal{I}$. Then $\mathcal{L}\mathcal{L} = \mathcal{I}$.

EXERCISES

1. Some of the following expressions stand for points, some for geometric transformations, and some are meaningless. Which are which? We agree that P and Q designate points and that $\mathcal{L}$ and $\mathcal{M}$ have the meanings just given them.

$$P\mathcal{M}\mathcal{M}\mathcal{L}; \quad P\mathcal{M}P\mathcal{L}; \quad \mathcal{L}\mathcal{M}\mathcal{L}\mathcal{L}; \quad \mathcal{L}\mathcal{M}\mathcal{L}P;$$
$$\mathcal{M}\mathcal{L}\mathcal{M}\mathcal{L}\mathcal{M}\mathcal{L}\mathcal{L}\mathcal{M}; \quad \mathcal{M}\mathcal{L}P\mathcal{M}\mathcal{L}\mathcal{M}Q.$$

2. Draw a pair of intersecting lines l and m, mark a point P not on either line, and construct each of the following points: $P\mathcal{L}$, $P\mathcal{M}$, $P\mathcal{L}\mathcal{M}$,

$P\mathfrak{M}\mathcal{L}$, $P\mathfrak{M}\mathcal{L}\mathcal{L}\mathfrak{M}$, $P\mathfrak{M}\mathcal{L}\mathfrak{M}\mathcal{L}$. In each case put into words a description of the construction indicated by the given expression.

3. Carry out the construction as in the preceding exercise but take the lines to be perpendicular and find the points $P\mathcal{L}\mathfrak{M}$, $P\mathcal{L}\mathfrak{M}\mathcal{L}\mathfrak{M}$, $P\mathcal{L}\mathfrak{M}\mathcal{L}\mathfrak{M}\mathcal{L}\mathfrak{M}$ and $P\mathcal{L}\mathfrak{M}\mathcal{L}\mathfrak{M}\mathcal{L}\mathfrak{M}\mathcal{L}\mathfrak{M}$.

4. Construct as many different points $P\mathcal{L}$, $P\mathcal{L}\mathfrak{M}$, $P\mathcal{L}\mathfrak{M}\mathcal{L}\mathfrak{M}$, etc., as you can when l and m intersect at an angle of 45°.

5. Reflect the vertices of an equilateral triangle successively in its *three* sides. (Do not attempt to get *all* the points so obtainable!) If you used only two sides and the vertices opposite them, how many points would you obtain and how would they be arranged?

6. Show that $P\mathcal{L}\mathfrak{M}\mathcal{L}\mathcal{L}\mathfrak{M}\mathcal{L}$ coincides with P. How can we obtain P from $P\mathfrak{M}\mathcal{L}\mathfrak{M}$ by successive reflections? What geometric transformation should be inserted in the parentheses in order that $\mathcal{L}\mathfrak{M}\mathcal{L}\mathfrak{M}$ (   ) shall equal $\mathcal{I}$, the identity transformation?

4. Rotations.

If the reader has carried out the preceding constructions with care, he may have been struck by the fact that the points constructed fall into two separate patterns. In Fig. 7:5 we see that the points P, P_2, P_4, and P_6 are all obtained by proceeding counterclockwise around C, the "spacing" between successive points appearing to be about the same at each step. Similarly, we get from P_1 to P_3 by proceeding clockwise around C and to P_5 from P_3 by continuing clockwise an equal amount.

That it is actually true that $\angle PCP_2 = \angle P_2CP_4 = \angle P_4CP_6 =$ etc., is a consequence of the following theorem.

Theorem. If P is any point other than C, the angle from CP to CP_2, where $P_2 = P\mathcal{L}\mathfrak{M}$, is twice the angle from line l to line m.[1]

[1] There are, of course, two positive angles less than 180° *between* the lines l and m. We shall agree that the phrase "the angle from line l to line m" is to mean the angle which, when described in the counterclockwise sense, has for its initial side l and its terminal side m. While this description will enable you to "mark in a figure" the correct angle, it is nevertheless true that it would be rather difficult to give a precise definition of "counterclockwise sense" without first giving rather careful proofs of a chain of theorems which you might consider too obvious to require proof! We also agree that the angle PCP_2, for instance, is the angle from PC to CP_2 in the above sense.

PROOF: In Fig. 7:7 we see that $\angle PCP_2 = \angle P_1CP_2 - \angle P_1CP$
$= \angle P_1CF_2 + \angle F_2CP_2 - (\angle P_1CF_1 + F_1CP)$. Since the triangles
P_1CF_1 and PCF_1 are congruent (why?), $\angle P_1CF_1 = \angle F_1CP$.
Similarly, $\angle P_1CF_2 = \angle F_2CP_2$ and hence $\angle PCP_2 = 2(\angle P_1CF_2)$
$-2(\angle P_1CF_1) = 2[(\angle P_1CF_2) - (\angle P_1CF_1)] = 2 \angle F_1CF_2 = 2\,\theta$. To
make our proof complete, we should carry it through when
P is in the "other" angle between the lines l and m, but we
shall leave this as an exercise for the student.

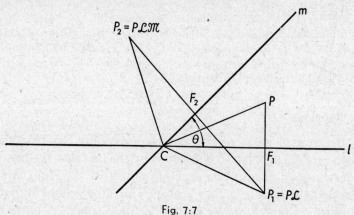

Fig. 7:7

With the aid of this theorem it is very easy to construct
as many of the points $P\mathcal{LM}$, $P\mathcal{LMLM}$, $P\mathcal{LMLMLM}$, etc., as
we wish. Thus in Fig. 7:8 we get the points marked by
making arcs PP_2, P_2P_4, P_4P_6, etc., all equal to twice the
arc $\mathcal{LM}$. ($P_2 = P\mathcal{LM}$, $P_4 = P\mathcal{LMLM}$, etc.)

One rather unexpected result follows from this construction. The points P_2, P_4, P_6, etc., would remain in precisely
the same position if instead of reflecting P in l and m we
were to reflect P successively in two lines l' and m' through C
*provided the angle from l' to m' is the same as the angle from
l to m.* This fact is illustrated in Fig. 7:9.

The geometric transformations $(\mathcal{LM})$ and $(\mathcal{L'M'})$ evidently
have precisely the same effect on every point in the plane.
This effect is briefly described by saying that the entire
plane is turned or *rotated* around C through an angle $2\,\theta$.
We agree to take for granted the additional phrase "in a

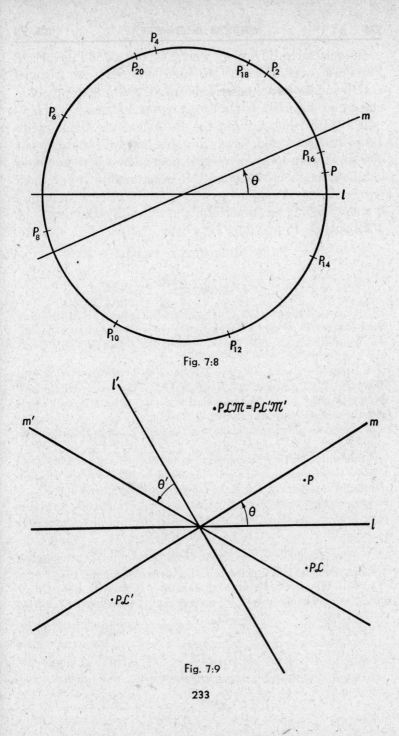

Fig. 7:8

• $P\mathcal{L}\mathcal{M} = P\mathcal{L}'\mathcal{M}'$

Fig. 7:9

233

counterclockwise direction" when θ is positive and the phrase "in a clockwise direction" when θ is negative.

If we write out a description of, say, P_{16} in terms of $\mathcal{L}$ and $\mathfrak{M}$, we have $P_{16} = P\mathcal{L}\mathfrak{M}\mathcal{L}\mathfrak{M}\mathcal{L}\mathfrak{M}\mathcal{L}\mathfrak{M}\mathcal{L}\mathfrak{M}\mathcal{L}\mathfrak{M}\mathcal{L}\mathfrak{M}\mathcal{L}\mathfrak{M}$. However, we can let $\mathfrak{R}$ stand for the *rotation* $\mathcal{L}\mathfrak{M}$ and write $P_{16} = P\mathfrak{R}\mathfrak{R}\mathfrak{R}\mathfrak{R}\mathfrak{R}\mathfrak{R}\mathfrak{R}\mathfrak{R}$. Even now we may still damage our eyesight trying to keep track of how many $\mathfrak{R}$'s there are in our description of the point. It is natural to count them once and for all and to write $P_{16} = P\mathfrak{R}^8$, the "exponent" 8 merely telling us to perform the rotation eight times. A point such as P_7 may now be written

$$P_7 = P\mathcal{L}\mathfrak{M}\mathcal{L}\mathfrak{M}\mathcal{L}\mathfrak{M}\mathcal{L} = P\mathfrak{R}^3\mathcal{L}.$$

EXERCISES

1. Carry out the proof of the above theorem when P is in the other angle between l and m. Compare your proof with that given and show that the above proof is valid for all positions of P (other than C) if clockwise angles are regarded as negative.

2. Using the notation of Fig. 7:7, show that $\angle PCQ = 360° - 2\theta$, where $Q = P\mathfrak{M}\mathcal{L}$. (Recall that the angle PCQ is the "counterclockwise" angle with initial side CP and terminal side CQ.) Does this result agree with the theorem proved in section 4? Explain.

3. Why does the theorem show that the angle between two successive points P_2, P_4, P_6, $\cdots$ is the same?

4. Show that if $\mathfrak{R}^{-1}$ is an abbreviation for $\mathfrak{M}\mathcal{L}$, then $\mathfrak{R}^{-1}$ is a *clockwise* rotation through an angle 2θ and that $\mathfrak{R}^{-1}\mathfrak{R} = \mathfrak{R}\mathfrak{R}^{-1} = \mathcal{I}$, the identity transformation. Letting $\mathfrak{R}^{-b}$ stand for the repetition b times of $\mathfrak{R}^{-1}(b$ an integer), and letting $\mathfrak{R}^0 = \mathcal{I}$, show that

$$\mathfrak{R}^a \cdot \mathfrak{R}^{-b} = \mathfrak{R}^{a-b} = \mathfrak{R}^{-b} \cdot \mathfrak{R}^a,$$
$$\mathfrak{R}^a\mathcal{L} = \mathcal{L}\mathfrak{R}^{-a}, \quad \mathfrak{M}\mathfrak{R}^a = \mathfrak{R}^{-a}\mathfrak{M},$$
$$\mathfrak{R}^a\mathfrak{M} = \mathfrak{R}^{a-1}\mathcal{L}, \text{ and } \mathcal{L}\mathfrak{R}^a = \mathfrak{R}^{-a+1}\mathfrak{M}.$$

Are these formulas still true if a is zero or a negative integer?

5. With the agreements as to notation made in Exercise 4, show that any point obtainable from P by a succession of reflections in l and m must occur in one of the lists:

$$P, \ P\mathfrak{R}, \ P\mathfrak{R}^2, \ P\mathfrak{R}^3, \ \cdots$$
$$P\mathcal{L}, \ P\mathfrak{R}\mathcal{L}, \ P\mathfrak{R}^2\mathcal{L}, \ P\mathfrak{R}^3\mathcal{L}, \ \cdots$$
$$P, \ P\mathfrak{R}^{-1}, \ P\mathfrak{R}^{-2}, \ P\mathfrak{R}^{-3}, \ \cdots$$
$$P\mathfrak{M}, \ P\mathfrak{R}^{-1}\mathfrak{M}, \ P\mathfrak{R}^{-2}\mathfrak{M}, \ P\mathfrak{R}^{-3}\mathfrak{M}, \ \cdots.$$

6. Show that if L is a line, and if all points on L are rotated through an angle of 2θ about a fixed point P, not necessarily on L, they will then be on a line L' making an angle of 2θ with L. (*Hint:* Use the theorem of the previous section, taking line l parallel to L.)

7. Show that if, in Exercise 6, θ is greater than $0°$ and less than $90°$, then there is only one point Q on line L which is taken by the rotation into another point on L. Call Q' this latter point and show (1) that the line segment QQ' is bisected by the foot of the perpendicular from P upon L; (2) that angle $QPQ' = 2\theta$, and (3) that Q' is the point of intersection of L and L'.

8. If, in Exercise 6, $\theta = 90°$, can there be a point Q as described in the previous exercise? If so, under what conditions? Whether or not Q exists, how will P be located with reference to lines L and L'?

5. Completion or exhaustion?

It is now rather natural to ask ourselves if we can construct the complete set of points obtainable from P by successive reflections in l and m, or if the number of points we can construct is limited only by the point at which we become too exhausted to continue the matter further.

Evidently we can obtain an arbitrarily large number of points P, $P\mathfrak{R}$, $P\mathfrak{R}^2$, $P\mathfrak{R}^3$, $\cdots$ unless some of these points coincide. For instance, $P\mathfrak{R}^2$ might coincide with $P\mathfrak{R}^{10}$. Then $P\mathfrak{R}^3$ would coincide with $P\mathfrak{R}^{11}$, $P\mathfrak{R}^4$ with $P\mathfrak{R}^{12}$, etc. But in order for $P\mathfrak{R}^2$ to coincide with $P\mathfrak{R}^{10}$ it would be necessary that $P\mathfrak{R}$ should coincide with $P\mathfrak{R}^9$ and that P should coincide with $P\mathfrak{R}^8$. Evidently if $P\mathfrak{R}^c$ coincides with $P\mathfrak{R}^d$, where d is greater than c, P must coincide with $P\mathfrak{R}^{d-c}$. Hence, if the number of points obtainable from P by successive rotations is to be finite, we must have $P = P\mathfrak{R}^b$ for some integer b. But this is possible only if the angle $b \cdot 2\theta$ through which P has been rotated to obtain $P\mathfrak{R}^b$ is an integral multiple of $360°$. That is, $b \cdot 2\theta = a \cdot 360°$ or $\theta = \frac{a}{b} \cdot 180°$, where a and b are integers.

Since we may draw the lines l and m so that θ will be any size we please,[1] we may choose to take $\theta = \frac{1}{\sqrt{2}} \cdot 180°$. Then

---

[1] That this is so depends upon the axiom of completeness of Euclidean geometry. See page 25 of reference **22**.

the sequence of points P, $P\Re$, $P\Re^2$, $\cdots$ will all be different for otherwise $180°(1/\sqrt{2}) = 180°(a/b)$ or $\sqrt{2} = b/a$ and we saw in Chapter III that it is impossible to express $\sqrt{2}$ as the quotient of two integers.

EXERCISES

1. How many rotations through an angle of 30° are necessary in order that every point shall return to its initial position? How many if the angle is 72°? 73°? $7\frac{1}{2}$°? $37\frac{1}{2}$°?

2. Show that if $\theta = 180°(a/b)$, then not more than $2\,b$ distinct points will be obtained by reflecting P in l and m. Will fewer points be obtained for some positions of P?

3. Show that if P is chosen so that the angle from CP to l equals θ, then $P\Re\mathcal{L} = P$. For what positions of P will $P\Re^b\mathcal{L} = P$?

4. If exactly $2\,k$ distinct points can be obtained by reflecting a point P in lines l and m, what can you say about the size of the angle between l and m?

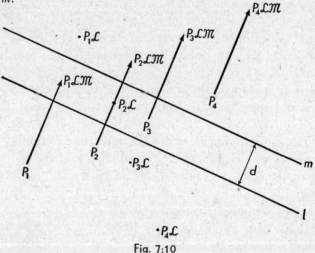

Fig. 7:10

6. Parallel mirrors.

Let us now determine the effect of successive reflections in parallel lines. In Fig. 7:10 we show the effect of reflecting some points in the parallel lines l and m. The arrows in the figure are drawn to call attention to the fact that the transformation $\mathcal{L}\mathfrak{M}$ moves every point a distance $2\,d$ in a direction perpendicular to l and m. We leave it as an exercise for the reader to prove that this is true for all positions of P.

From the preceding statement it evidently follows that the transformation which carries P into $P\mathcal{L}\mathfrak{M}$ may equally well be determined by lines l' and m' parallel to l and m and equally far apart.[1] Thus in Fig. 7:11, $P\mathcal{L}$ and $P\mathcal{L}'$ are different but $P\mathcal{L}\mathfrak{M} = P\mathcal{L}'\mathfrak{M}'$.

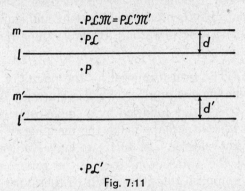

Fig. 7:11

The transformation $\mathcal{L}\mathfrak{M}$ illustrated in Fig. 7:10 is determined by any one of the parallel and equal arrows of that figure. We shall call such a transformation a **translation** and denote it by T.

EXERCISES

1. Prove that if P is reflected in line l and then in line m, parallel to l, P goes into a point whose distance from P is $2\,d$ where d is the distance from l to m.

2. Describe what you mean by a "translation."

3. Given the two translations T_1 and T_2, let $T_1 \cdot T_2$ (or $T_1 + T_2$) denote the result of performing first T_1 and then T_2. Show that $T_1 \cdot T_2$ is a translation. Are $T_1 \cdot T_2$ and $T_2 \cdot T_1$ the same translations?

4. How can one obtain from T the translation T^{-1} having the property that $TT^{-1} = \mathcal{I}$? What is T^{-1} in terms of $\mathcal{L}$ and $\mathfrak{M}$ when $T = \mathcal{L}\mathfrak{M}$ and $\mathcal{L}$ and $\mathfrak{M}$ are reflections in parallel lines l and m. Is $T^{-1}T = \mathcal{I}$?

[1] We must also require that m' be on the same side of l' as m is of l. This intuitively evident requirement may be accomplished by demanding that the distance between l and l' shall equal the distance between m and m'.

7. Many mirrors.

An alert reader is now likely to ask what happens to a point when it is successively reflected in three or more lines. Does the combination of three reflections in lines result in a hopelessly confused "shuffling" of the points in the plane, or is it possible to give a simple description of the resulting transformation? Actually the latter is the case but a complete proof (by the methods we have been using) of the following plausible theorem would be long and clumsy. (The proof would be easy if we had available some results from analytic geometry.)

Theorem. The transformation of the points of the plane resulting from an *even* number of reflections may be effected by a single translation followed by a single rotation.

Figure 7:12 will perhaps make the meaning of this theorem somewhat clearer. In fact, we have indicated in it how to find the translation, T, and the rotation, $\mathcal{R}$, necessary to accomplish the same result as the four reflections: $T\mathcal{R} = klmn$. In the figure A and B are chosen as any two points. We first construct $Aklmn$ and $Bklmn$. Then we take T to be the translation which carries A into $Aklmn$ and $\mathcal{R}$ to be the rotation with center $Aklmn$ which carries BT into $Bklmn$. Our claim is that $CT\mathcal{R} = Cklmn$ for any other point C.

Without attempting a formal proof let us see why this is so in terms of Fig. 7:12. By Exercise 6 of section 4, the angle from line AB to $A'B'$ (where $A' = Akl$ and $B' = Bkl$) is twice the angle, θ_1, from line k to line l; also the angle from $A'B'$ to $A''B''$ (where $A'' = Aklmn$ and $B'' = Bklmn$) is twice the angle from line m to line n. If you then draw the three lines AB, $A'B'$, $A''B''$ so that all the intersections show and use the plane geometry theorem that an exterior angle of a triangle is equal to the sum of the opposite interior angles, you will see that the angle from AB to $A''B''$ is $2(\theta_1 + \theta_2)$. The same will be true of AC and $A''C''$ and BC and $B''C''$. Since lengths are left unaltered by reflections, we see that the effect of the four reflections is to perform the translation on the triangle ABC indicated in Fig. 7:12 followed by rotation through an angle $\theta = 2(\theta_1 + \theta_2)$.

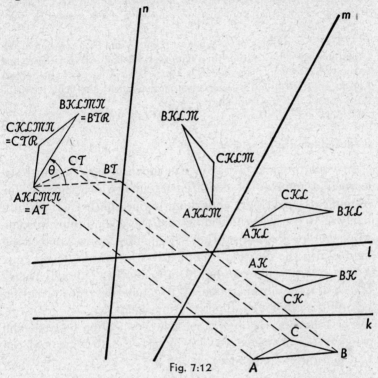

Fig. 7:12

To describe the effect of an odd number of reflections, we use the fact that any odd number is an even number (or zero!) plus one. Hence, the result of an odd number of reflections may always be obtained by successively performing a single translation, a single rotation, and a single reflection. In symbols, $\mathcal{L}_1 \cdot \mathcal{L}_2 \cdot \mathcal{L}_3 \cdots \mathcal{L}_{2n} \cdot \mathcal{L}_{2n+1} = T\mathcal{R}\mathcal{L}_{2n+1}$ where $\mathcal{L}_1 \cdot \mathcal{L}_2 \cdots \mathcal{L}_{2n} = T\mathcal{R}$.

EXERCISES

1. Draw a figure corresponding to Fig. 7:12 for different lines k, l, m, n.
2. Show how to find two lines such that successive reflection in them will result in a given translation.

3. Show how to find two lines such that successive reflection in them will result in a rotation through a given angle about a given point.

* 4. Given two line segments AB and $A'B'$ of equal length, show how to find four lines k, l, m, and n such that $A\mathcal{KLMN} = A'$ and $B\mathcal{KLMN} = B'$. How is your construction of the lines effected if AB is parallel to $A'B'$? if A' coincides with A?

* 5. If $\mathcal{R}_1$ is a rotation of 60° about a point C_1 and $\mathcal{R}_2$ a rotation of 60° about another point C_2, show in a figure how to find a translation T and a rotation $\mathcal{R}$ such that $\mathcal{R}_1 \cdot \mathcal{R}_2 = T\mathcal{R}$. (*Hint:* Construct the points $C_1\mathcal{R}_1\mathcal{R}_2$ and $C_2\mathcal{R}_1\mathcal{R}_2$.) Do the transformations $\mathcal{R}_1\mathcal{R}_2$ and $\mathcal{R}_2\mathcal{R}_1$ have the same effect on all the points in the plane?

* 6. Solve Example 5 when the angles are both 90°.

8. Euclidean displacements.

In your previous work on Euclidean geometry you have learned a number of conditions under which figures (for example, triangles) are **congruent** to one another. It is quite likely that you had the feeling that these conditions in some way or other guaranteed that you might "move" one of the figures into the position occupied by the other. This vague notion of "moving one figure into another" should be replaced by the sharper one of a geometric transformation before it is employed directly in mathematics.

The following theorem states the connection between the geometric transformations we have been studying and the relationship of congruence.

Theorem. Two figures are congruent if and only if one of them may be carried into the other by a succession of reflections in straight lines.

The proof of the theorem is too difficult to give here, but you will probably regard it as plausible in view of the possibility of obtaining arbitrary translations and rotations by the successive performance of reflections (see exercises 2 and 3 above).

The set of all the geometric transformations obtainable by the successive performance of reflections is sometimes called the **group** [1] **of Euclidean displacements.** In recent

---

[1] This term was defined in section 11 of Chapter II.

years mathematicians have come to regard the group of displacements as fundamental and to use it to define other concepts. Thus, instead of beginning the study of Euclidean geometry by writing down as agreed upon certain axioms establishing properties of points, lines, congruent angles, etc., we might begin by describing in some convenient fashion [1] a class of points and a group of geometric transformations on them. Then we should be free to *define* two figures as congruent when some one of our transformations carries one into the other.

[1] For example, by using coordinates for the points and equations to describe the geometric transformations. An example of this kind is given in the next chapter.

CHAPTER VIII

Lorentz Geometry

"I don't know what you mean by 'glory,'" Alice said.

Humpty Dumpty smiled contemptuously. "Of course you don't — till I tell you. I meant 'there's a nice knockdown argument for you.'"

"But 'glory' doesn't mean 'a nice knockdown argument,'" Alice objected.

"When *I* use a word," Humpty Dumpty said in rather a scornful tone, "it means just what I choose it to mean — neither more nor less."

"The question is," said Alice, "whether you *can* make words mean so many different things."

"The question is," said Humpty Dumpty, "which is to be master — that's all!"

1. Introduction.

The usual approach to Euclidean geometry is to write down at the very beginning certain statements called **axioms**. These axioms are in many ways analogous to the rules of such a game as checkers or chess. Their great age and the general familiarity of mankind with them has won these axioms a veneration which is not their due. Indeed, they are not even so very old — they are younger than the Pyramids — and every teacher of geometry will agree that not all persons are familiar with them.

Even a checker game could hardly proceed smoothly if at every play one of the players insisted on a change in the rules of the game. In just the same way the axioms of Euclidean geometry, *once accepted at the start of the theory*, cannot be contradicted at any later stage in its development.

If, however, the players were to grow tired of their game of checkers and begin to play chess, we should not be surprised to see them making moves quite impossible before. In just the same way mathematicians sometimes put away their perpendicular lines, their circles, and their equilateral triangles and play the game of a new and different kind of geometry. They even draw diagrams on a sheet of paper to describe their new game, just as one plays chess on a checkerboard.

The historically important approach to a geometry different from that of Euclid is to modify one or more of the Euclidean axioms and study the consequences of the new rules. This was done in the early nineteenth century by three men, Gauss, Bolyai, and Lobachevsky, working independently of one another. The general recognition that there were other geometric "games" to be played had far-reaching effects in all domains of human thought.

We shall not hope to acquaint the student with this sort of "non-Euclidean" geometry in these pages. (See section 7.) Instead, we shall attempt to describe very briefly the foundations of a geometry which is different from both Euclid's and that frequently discussed under the title Non-Euclidean Geometry. In some respects our theory is easier to formulate than either of the historically more important types. It may also lay claim to being of greater interest in contemporary physics, although this aspect of the subject we must disregard entirely.

2. Lorotations.

At the close of the last chapter we pointed out the possibility of using the groups of displacements of Euclidean geometry to define what is meant by congruent figures (in the Euclidean sense of "congruent"). There it would have been fairly difficult to give a description of the displacements by means of equations connecting the coordinates of the original and the transformed point. In our new geometry this is much the simplest way of describing the geometric transformations which carry a figure into (what is *defined* to be) a **congruent** figure.

Let us then agree to play our new game on the same "board" that we used in Chapter V. By this we mean that a pair of real numbers (x,y) will determine a **point**, a **straight line** will be the set of all points determined by solutions of a linear equation $Ax + By + C = 0$ (with not both A and B zero), and a pair of linear equations will determine **parallel lines** (definition!) if and only if the equations have no common solution.

We have as yet no way of "displacing" one figure into another. If we were to use the rotations, translations, and reflections of the last chapter for this purpose, we should, of course, be studying Euclidean geometry. Instead, we agree that the transformations which carry a figure into a "congruent" figure are precisely those which carry the arbitrary point (x,y) into the point (x',y'), where

$$x' = kx + c,$$
$$y' = y/k + d,$$

with c and d *any* real numbers and k any *positive* real number.

To obtain a little more insight into what this transformation "does" to the points of the plane, let us take $c = 2$,

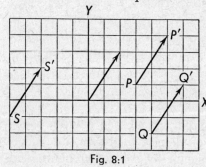
Fig. 8:1

$d = 3$, and $k = 1$. Then P, Q, and S go into P', Q', and S', respectively, as indicated in Fig. 8:1.

The arrows suggest that the transformation is our old friend a translation. This is, indeed, the case *whenever $k = 1$*. The proof is left as an exercise.

If, on the other hand, we take k different from 1, say $k = 2$, and set $c = d = 0$, we obtain the transformation $x' = 2\,x$, $y' = \frac{1}{2}\,y$. Corresponding points under this transformation are marked in Fig. 8:2 (P_1 goes into P_1', P_2 goes into P_2', etc.).

P	(1,0)	(3,0)	(1,1)	(−4,−1)	(−3,−2)	(0,4)
P'	(2,0)	(6,0)	(2,$\frac{1}{2}$)	(−8,−$\frac{1}{2}$)	(−6,−1)	(0,2)

The transformations with $c = d = 0$ leave the point (0,0) unmoved and are in some respects analogous to Euclidean

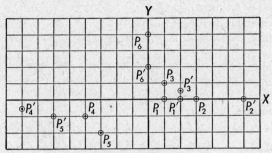

Fig. 8:2

rotations. Let us call them **lorotations**.[1] For simplicity's sake, we shall in the remainder of this chapter take $c = d = 0$ and consider only the effect of lorotations on figures.

EXERCISES

In each of the following problems plot the given points and the points after they are transformed by the given lorotation. Draw a smooth curve through the original points and another smooth curve through the transformed points. Where the curves appear to be straight lines, find their equations.

1. Points: (1,1), (2,1), (3,1), (4,1), (5,1), (6,1).
   Lorotation: $x' = \frac{1}{3} x$, $y' = 3 y$.

2. Points: (2,1), (4,2), (6,3), (8,4), (−4,−2).
   Lorotation: $x' = \frac{1}{2} x$, $y' = 2 y$.

3. Points: (2,0), (4,1), (6,2), (8,3), (−4,−3).
   Lorotation: $x' = \frac{1}{2} x$, $y' = 2 y$.

4. Points: (1,1), (3,−3), (−2,7), (0,3).
   Lorotation: $x' = 2 x$, $y' = \frac{1}{2} y$.

5. Points: (5,0), (5,1), (5,4), (5,−2).
   Lorotation: $x' = \frac{1}{5} x$, $y' = 5 y$.

6. Points: (13,0), (12,5), (5,12), (0,13), (−5,12), (−12,5), (−13,0), (−12,−5), (−5,−12), (0,−13), (5,−12), (12,−5).
   Lorotation: $x' = \frac{1}{2} x$, $y' = 2 y$.

[1] Lorotation = lo (an expression of wonderment) + rotation! More seriously, the geometry we are studying is a very much simplified version of the sort of geometry developed by *Lorentz* and now of importance in the theory of relativity.

7. Points: (5,0), (8,1), (10,5), (9,8), (5,10), (2,9), (0,5), (1,2).
   Lorotation: $x' = \frac{1}{3}x$, $y' = 3y$.

8. Points: $(\frac{5}{2},0)$, (2,6), $(\frac{3}{2},8)$, (0,10), $(-2,6)$, $(-\frac{5}{2},0)$, $(-\frac{3}{2},-2)$, $(0,-10)$, $(2,-6)$.
   Lorotation: $x' = 2x$, $y' = \frac{1}{2}y$.

9. Points: (6,1), (3,2), (2,3), (1,6).
   Lorotation: $x' = \frac{3}{2}x$, $y' = \frac{2}{3}y$.

10. Points: (12,1), (6,2), (4,3), (3,4), (2,6), (1,12).
    Lorotation: $x' = 2x$, $y' = \frac{1}{2}y$.

11. Points: (14,7), (8,8), (6,9), (5,10), (4,12), (3,8).
    Lorotation: $x' = \frac{1}{2}x$, $y' = 2y$.

12. Points: (1,13), (2,8), (3,7), (4,7), (6,8), (12,13).
    Lorotation: $x' = 2x$, $y' = \frac{1}{2}y$.

13. Into what point do the lorotations $x = kx'$, $y = \frac{1}{k}y'$ take the point (2,3) when $k = 2, 3, 4$, and 5. Draw the graphs of these points.

14. The line connecting $P_1 : (2,3)$ and $P_2 : (4,6)$ goes through the origin. Can the same thing be said of the points P_1' and P_2' where P_1' and P_2' are the points obtained from P_1 and P_2 by the lorotation $x = kx'$, $y = \frac{1}{k}y'$?

15. Prove that all the points on a line through the origin are transformed by a lorotation into points on another (or the same) line through the origin.

* 16. Prove that any set of points on a line is transformed by a lorotation into a set of points on another (or the same) line.

* 17. Prove that in Exercise 16 both lines are the same if k in the lorotation is $=1$ and only in that case.

3. Models.

The fact that Euclidean geometry was invented to describe our physical surroundings makes it very easy to construct models of our figures in this kind of geometry. Thus a long thin rod or a broad "straight" carbon mark on a piece of paper may be used to suggest the abstract concept "straight line." We can suggest the effect of a displacement on a figure by drawing the figure on a sheet of paper and then "moving" the sheet around (perhaps turning it over). Of course, if we had happened to draw our figure on a sheet of rubber, we should have had to be very careful not to stretch the rubber in any way, for, if we did, our model of the geometry would not be useful.

It cannot be emphasized too strongly that the use of any model at all in geometry is not only unnecessary, it is positively dangerous *unless we keep clearly in mind at all times that the model is merely a convenient way of describing the abstract system of relationships which constitute the geometry.*

It is very difficult to construct any satisfactory model of our new geometry, chiefly because a lorotation is such a queer "motion." We notice, however, that the lorotation $x' = kx$, $y' = \frac{1}{k} y$, multiplies all x coordinates by the same factor.

Since $0 \cdot k = 0$, a point on the y-axis is moved into another point on the y-axis. The points on a line $x = 3$ are all moved into points with x coordinate $3 k$ and are, therefore, all on the line $x = 3 k$. Similarly, the line $x = 7$ goes into the line $x = 7 k$ and the line $x = -4$ goes into the line $x = -4 k$. Evidently the effect of our lorotation is to "stretch" the plane in the x-direction, keeping the y-axis fixed and moving all lines parallel to it in such a way that their distance from the y-axis is multiplied by k. If k is less than 1, this would be described as a contraction.

From similar considerations we see that the x-axis stays fixed under a lorotation and that lines parallel to it are moved so that their distances from it are multiplied by $1/k$.

A model of our geometry, which we do not seriously propose to use, would be a curtain on a curtain stretcher. Of course, when we stretched the curtain in length we should have to allow it to contract in width — not a recommended procedure in drying curtains.

4. Common properties.

Let us see if we can find any theorems of Euclidean geometry which continue to hold in our new geometry. Looking through the first few pages of Hilbert's *Foundations of Geometry* (reference **22** in the Bibliography), we see the following theorems which could still be proved in our system. (The proofs are not too difficult to be worked out by a student who is well versed in algebra.)

Two distinct points A and B always completely determine a straight line.[1] We, in effect, proved this in Chapter V when we showed how to find the equation of a line through two given points.

Any two distinct points of a straight line completely determine that line; that is, if A and B determine a line l and A and C determine the same line l, where B and C are different, then B and C determine the line l.[1]

Two distinct straight lines have either one point or no point in common.[1]

Through a point not on a line l there can be drawn one and only one straight line which does not intersect the line l.[1] (We have agreed to call two such lines **parallel**.)

If two straight lines are parallel to a third, then they are parallel to each other.

5. Uncommon properties.

None of the theorems just stated made any reference to the distance between two points. This is because a lorotation does not preserve what we call in Euclidean geometry the *distance* between two points. For instance, the lorotation $x' = 4x$, $y' = y/4$ carries $(0,0)$ into $(0,0)$ and $(3,4)$ into

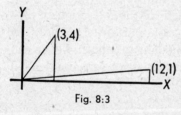

Fig. 8:3

$(12,1)$. In Fig. 8:3 we see that the *Euclidean distance* from $(0,0)$ to $(3,4)$ is 5, while the distance between the transformed points $(0,0)$ and $(12,1)$ is $\sqrt{145}$, which is not equal to 5. Notice, however, that the areas of the two triangles are equal. (This latter would not be true if the point $(3,4)$ were rotated through an angle of $30°$, for instance.)

Having observed this fact we shall not be surprised if the points on a "circle" about the origin are moved by a loro-

[1] These statements are taken as axioms in Euclidean geometry. Since we did not take them as basic assumptions, we are obliged to prove them as consequences of our assumptions.

tation to positions off the circle. Thus the points (0,5), (3,4), (4,3), and (5,0) are all five Euclidean units of distance from the origin, but the lorotation $x' = 2\,x$, $y' = \frac{1}{2}\,y$ moves them to the positions shown in Fig. 8:4. The points on the "circle" C will in fact all go into points on the curve C'.

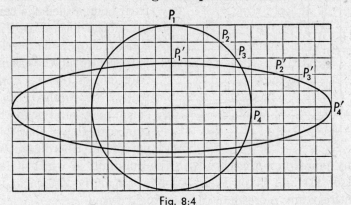

Fig. 8:4

The fundamental property of a "circle" in Euclidean geometry is that we can get as many points of it as we please by rotating about the center through different angles any one point on it. Let us see if we can discover a curve having an analogous property under lorotations. We start with the point with coordinates (3,2) and perform the following lorotations:

$$x' = \tfrac{1}{3}\,x,\ y' = 3\,y \quad \text{transforms (3,2) into (1,6)}$$
$$x' = \tfrac{1}{2}\,x,\ y' = 2\,y \quad \text{transforms (3,2) into } (\tfrac{3}{2},4)$$
$$x' = 2\,x,\ y' = \tfrac{1}{2}\,y \quad \text{transforms (3,2) into (6,1)}$$
$$x' = 3\,x,\ y' = \tfrac{1}{3}\,y \quad \text{transforms (3,2) into } (9,\tfrac{2}{3})$$

Drawing the graphs of all these points and a smooth curve connecting them, we get Fig. 8:5. The equation $xy = 6$ is evidently satisfied by all the points of the form $(k \cdot 3,\ 2/k)$ and hence every point obtained from (3,2) by a lorotation will lie on this curve. However, it is only that part of the curve $xy = 6$ in the first quadrant which we can obtain by "lorotating" (3,2). (We leave it to the student to draw the rest of the locus of $xy = 6$.)

Let us call a curve such as that in Fig. 8:5 a **lorcle** (pronounced: "lorkle"). Formally, we define a lorcle to be the

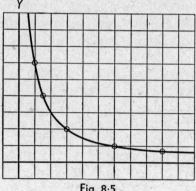

locus of all points obtainable from a single point
by lorotations about a
fixed point called the **center** of the lorcle. In Fig.
8:6 we show some lorcles
with center at the origin.
Any one of them is com-
pletely determined by a
single point on it or by the
common value of the prod-
uct of the coordinates of

Fig. 8:5

any point on it together with the quadrant in which it lies
(I or III and II or IV). A lorcle cannot be drawn in its
entirety any more than can a straight line.

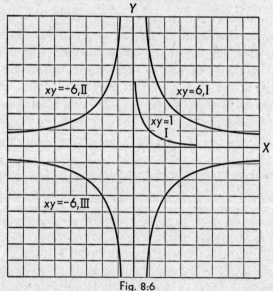

Fig. 8:6

In Euclidean geometry all except one of the lines through
a point on the circumference of a circle cut it in two points.
The one exceptional line touches the circle in only one point

and is called the **tangent** to the circle at the point. Is it possible to determine a line tangent to a lorcle in the same way? We shall do this for one numerical example, leaving it to any interested student to carry out the construction for any point on a lorcle with center at the origin.

Let the lorcle be the one shown in Fig. 8:5. It consists of all the points (x,y) which satisfy the equation $xy = 6$ and lie in the first quadrant. To find a tangent line at the point $(3,2)$ notice that a line $y = ax + b$ will pass through the point $(3,2)$ if $2 = 3a + b$; that is, if $b = 2 - 3a$. To determine the other point in which the line $y = ax + (2 - 3a)$ cuts the lorcle $xy = 6$, we must solve the equations simultaneously. Substituting in $xy = 6$ we get

$$x[ax + (2 - 3a)] = 6,$$

or

$$ax^2 + (2 - 3a)x - 6 = 0.$$

Factoring, we get

$$(ax + 2)(x - 3) = 0.$$

Hence $x = 3$ or $-2/a$. The line, therefore, intersects the curve $xy = 6$ in the points $(3,2)$ and $(-2/a, -3a)$. These points are distinct unless $3 = -2/a$, or $a = -2/3$. By analogy with Euclidean geometry we call this line (when $a = -2/3$) with equation

$$y = -2x/3 + 4,$$

the **tangent to the lorcle** through the point $(3,2)$.

We have already seen that the Euclidean distance between two points does not remain the same under lorotations. This might make us suspect that our usual notions of angle need to be abandoned. That this is, indeed, the case we can see as follows.

Let us refer to any figure made up of two half-lines emanating from the same point as an "angle." We must carefully refrain from associating any numerical measure with such a figure for the measurement of an angle in degrees requires a whole sequence of moves in the game of Euclidean geometry, and we are by no means sure that we can make analogous moves in the game we are playing.

A particular angle consists in the "positive" half of the x-axis and the half-line from the origin through the point $(1,1)$. The lorotation $x' = 2\,x$, $y' = \frac{1}{2}\,y$ carries the origin and the positive half of the x-axis into themselves, and the line OB into the line OC through $(2,1/2)$. (See Fig. 8:7.) Hence the angles AOB and AOC are congruent!

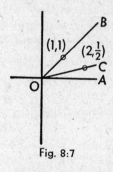

Fig. 8:7

This astonishing result should convince the student that the measurement of angles in our new geometry differs radically from the measurement of them in Euclidean geometry. The problem, while not insoluble, is too difficult for us here.

In spite of these difficulties, we can define perpendicular lines in our new system (or should we call them, say, **lopendicular?**). We do it by saying that the tangent to a lorcle is "perpendicular" to the line connecting the center with the point of contact. (What is the analogous theorem in Euclidean geometry?) For instance, by referring to our above discussion, we see that the line $y = -\frac{2}{3}\,x + 4$ is "perpendicular" to the line through $(0,0)$ and $(3,2)$. (What is the equation of the latter line?)

EXERCISES

1. If $P : (4,5)$ is carried by a lorotation into a point P', show that the area of the rectangle two of whose sides lie along the x and y axes and which has P as a vertex is the same as the area of the corresponding rectangle having P' as a vertex. Would a similar statement be true for any other point P?

2. If P is a point not on the x-axis, Q is the foot of the perpendicular from P upon the x-axis, and P' and Q' are obtained from P and Q by a lorotation, show that Q' is the foot of the perpendicular from P' upon the x-axis and that the area of the triangle PQO is the same as the area of triangle $P'Q'O'$, 0 being the origin.

3. Show that the line $y = mx - 2\,m + 4$ passes through $(2,4)$ for any value of the constant m. Find, in terms of m, the points in which the line intersects the curve $xy = 8$. For what value of b will these points

coincide? Draw the curve $xy = 8$ and the lines obtained by taking $m = -\frac{1}{2}$, $m = -1$, $m = -2$, and $m = -3$.

4. Find the perpendicular (in our new sense) at (5,3) to the line joining (5,3) to the origin.

* 5. Prove that if P_1 and P_2 are two points and 0 is the origin, then the area of the triangle P_1OP_2 is the same as the area of the triangle $P'_1OP'_2$ where P'_1 and P'_2 are obtained from P_1 and P_2 by a lorotation.

* 6. What is the equation of the line obtained by lorotating the line $y = -2\,x/3 + 4$ by the lorotation $x = kx'$, $y = y'/k$. Is the resulting line tangent to the lorcle $xy = 6$?

* 7. Show that the lorotations about the origin form a *group* in the sense in which this term was used in section 11 of Chapter II. (See reference **8**, pp. 539 ff.)

* 8. Show that the transformations

$$x' = x + c, \quad y' = y + d,$$

where c and d are arbitrary real numbers, form a group.

* 9. Show that the transformations

$$x' = kx + c, \quad y' = \frac{1}{k}y + d,$$

where c and d are arbitrary real numbers and k is positive, form a group.

6. Summary.

We could go on almost indefinitely making new definitions and proving new theorems in this geometry. If the student has caught the spirit of what we have already done, any additional proofs would be superfluous. The end we had in view was not so much to develop a body of theorems as to illustrate the possibility of studying mathematical systems distinctly different from Euclidean geometry, yet sufficiently similar to it to be called **geometries.**

In the next chapter we return to the familiar surroundings of Euclidean space, but discuss problems which must be attacked by methods not at all like those useful in a first course in Euclidean geometry.

7. Topics for further study.

1. Axioms of geometry: Reference **11**, pp. 239–250.

2. Geometry and art: Reference **8**, pp. 151–159; also almost any number of the periodical *Scripta Mathematica*.

3. Two- and four-dimensional geometry: References **25**, pp. 112–131; **1.**

4. Non-Euclidean geometry: See references **8**, Chap. XVIII; **11**, Chap. XI; **27**; **25**, pp. 131–155; **31**, Chap. XVI.

5. Foundations of Geometry: Reference **36.**

CHAPTER IX

Topology

---

1. The Königsberg bridge problem.

The city of Königsberg is situated on and around an island at the junction of two rivers. At the beginning of the eighteenth century the various parts of the city were connected by seven bridges, as shown in Fig. 9:1. The mathematicians of those days were interested in the following problem: Is it possible for a Königsberger to take a walk through the city during which he shall cross each of the seven bridges exactly once? Although a few trials will soon convince you that the answer is "no," it is not easy to prove this by trial, for there is a large number of ways of starting the walk. The proof was first given in 1736 by the famous mathematician, Euler.

A similar problem that is seen in puzzle books is to determine whether or not one can make a tour of the house in Fig. 9:2, going through each door once and only once and returning to the same room where the tour began. Here again the answer is "no," and due to the large number of doors it is very difficult to show this by trial.

The figures for these problems are similar in that they consist of several regions (districts of the city, rooms) connected by passageways (bridges, doors). We can simplify the figures considerably by replacing each region by a single point and the passageway between two regions by a line joining the corresponding points. In order to follow the transition more easily we label the regions (points) with numbers and the passageways (lines) with letters. If this is done, we have Fig. 9:3. (Notice that in the second example we must have a point labeled 1 corresponding with the region outside the house.) Our two original problems then, in terms of Fig. 9:3, are to traverse such a figure without

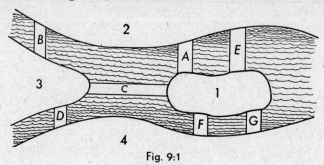

Fig. 9:1

covering any line more than once — or to draw the figure without lifting the pencil from the paper and without retracing any line. Other figures are given in Fig. 9:4.

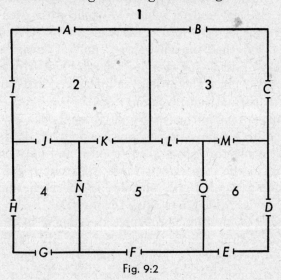

Fig. 9:2

Figures of this sort, consisting of a certain number of points joined in pairs by lines, no two of which intersect except at a vertex, shall be called **networks**. We shall call the points **vertices** and the lines **arcs**. A vertex is said to be **odd** or **even** according as an odd or an even number of arcs touch it. For example, the Königsberg network has 4 odd and no even vertices; network f in Fig. 9:4 has 14 even and 2 odd vertices.

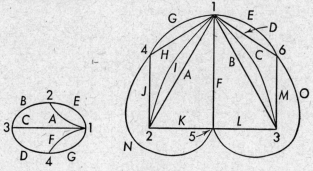

Fig. 9:3

EXERCISES

1. List the numbers of odd and even vertices in each of the networks of Figs. 9:3 and 9:4. Compare the number of odd vertices with your ability to draw the figure in one stroke of the pencil. Can you draw any conclusion from this comparison? If so, test the accuracy of your conclusion with other networks.

2. A wire framework is to be made in the form of the edges of a cube. What is the least number of pieces of wire required? Give reasons for your answer.

2. Paths in a network.

By a **path** in a network we shall mean a succession of arcs which can be traversed in a continuous manner without retracing any arc. A path is said to be **closed** if it begins and ends at the same vertex. A path is permitted to cross itself at one or more vertices.

It is evident that a network cannot be entirely traversed by a single path if it consists of two or more disconnected pieces. Hence, we will exclude such networks from our discussion, considering only those networks which are "connected." More precisely, a network is said to be **connected** if, for each pair of points A and B on the network, there is a path, in the above sense, joining A to B. By experimenting with various connected networks we may arrive at the following conclusions:

1. *The number of odd vertices in any network is even.* (Remember that zero is considered to be an even number.)

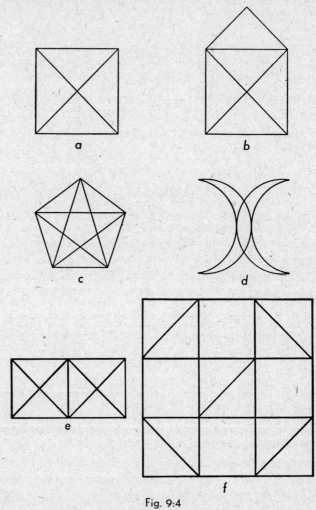

Fig. 9:4

2. *If a network has no odd vertices it can be traversed in a single closed path, starting at any vertex of the network and along any arc from that vertex.*

3. *If a network has two odd vertices it can be traversed in a single path, starting at one of the odd vertices and ending at the other.*

4. *If a network has more than two odd vertices it cannot be traversed in a single path.*

These are the statements which were proved by Euler. Applying statement 4 to the Königsberg bridge problem, we see that, since there are four odd vertices, the network cannot be traversed in a single path.

The proof of these four statements depends largely on the fact that if a is an even number, then $b - a$ is odd if b is odd and even if b is even. Here are the proofs.

1. The number of odd vertices in any network is even.

PROOF: Let n_1 be the number of vertices of the network which touch just one arc, let n_2 be the number of vertices which touch just two arcs, and so on. Then the number of odd vertices is

$$N = n_1 + n_3 + n_5 + n_7 + \cdots.$$

Now each arc has two ends, so the number of arc-ends is even. The n_1 vertices account for n_1 of these arc-ends, the n_2 vertices account for $2\,n_2$, etc. Hence,

$$S = n_1 + 2\,n_2 + 3\,n_3 + 4\,n_4 + 5\,n_5 + 6\,n_6 + 7\,n_7 + \cdots,$$

the total number of arc-ends, is even. Obviously,

$$S - N = 2\,n_2 + 2\,n_3 + 4\,n_4 + 4\,n_5 + \cdots$$

is an even number (and when *we* say "obviously" we mean "obviously"). Since S is even, N is also even, which is what we desired to prove.

2. If a network has no odd vertices it can be traversed in a single closed path starting at any vertex of the network and along any arc from that vertex.

PROOF: Suppose that our network has no odd vertices. Starting at any vertex, which we may call A, and along any arc from that vertex, we proceed in a path along the arcs. Suppose B is some vertex other than A. Since B is an even vertex, if we enter B by one arc there will be another arc by which we can leave. Moreover, having used up two arcs at B, the number left is still even, and so if we later return to B, the same situation will hold. Hence, we can always continue our path until we come back to A. We shall then have a closed path, which, of course, may not traverse the entire network. If it does, we are through; if it does not, we start over again with the unused portion of the network. This portion will still have all even vertices, since our closed path always accounted for the arc-ends two at a time. So the process may be continued, and we finally get a number of closed paths which among them use up all the arcs. Our problem now is to connect these closed paths into a single closed path. This is done as follows. If our first path does not traverse the whole network, there must be at least one of the other paths which has a vertex in common with it since the network itself is connected. Suppose the first path begins and ends at A, the second begins and ends at B, and a common vertex is C. Then we can combine the two paths into one by going from A to C, then around the second path from C to B and back to C, and along the rest of the first path to A. If there are any more paths left, the process can be continued until finally we have only one path beginning and ending with A and traversing the whole network.

3. If a network has two odd vertices it can be traversed in a single path, starting at one of the odd vertices and ending at the other.

PROOF: Suppose the network has exactly two odd vertices, A and B. Draw a new arc joining A and B. Then the new network will have no odd vertices, and so it can be traversed by a path starting and ending at B. By statement 2 above the path can be chosen so that the first arc traversed is the extra one from B to A. Now remove the extra arc, and we have a path starting at A and ending at B.

4. If a network has more than two odd vertices, it cannot be traversed in a single path.

PROOF: If a network can be traversed in a single path, no vertex, except possibly the ends of the path, can be odd.

For in going into and out of a vertex the arcs which it touches are used up in pairs, and so we can never use up an odd number. Since a single path has only two ends, such a network can have at most two odd vertices.

A more specific statement of 4 is

5. *If a network has 2 n odd vertices, n being a positive integer, it can be traversed in precisely n paths and in no less than n paths; in other words, the pencil must be lifted n − 1 times between the beginning of the path and the ending.*

The proof of this is left as an exercise.

EXERCISES

1. Review the exercises of the last section in the light of the results of this section.

2. Prove statement 5 above.

3. The four-color problem.

There is another famous problem which leads to the consideration of networks, and, like the Königsberg bridge problem, it concerns geography. A cartographer is making a map of a country which is divided into several districts. He wishes to color each district according to the usual convention that no two districts with a common border are to have the same color. How many different colors will he need?

To reduce this to a problem in networks we take a vertex for each of the districts and join two vertices by an arc whenever the corresponding districts have a common border. We then ask how many colors will be needed to color the vertices so that no arc will have the same color at both ends.

Fig. 9:5

Here, in Fig. 9:5, is the network corresponding to the New England states, neglecting the region outside New England.

The big difference between this question and the ones we asked previously about networks is that this one has never been answered. It has been proved that five colors are enough, and that three are not enough, but whether four will suffice in all cases is still undecided. In any particular network that has ever been investigated four colors have turned out to be sufficient, but this gives us no assurance that there might not be some network, say with several million vertices, which will require five colors. To settle the question one must either give an example of a network which cannot be colored with four colors, or else give a systematic method by means of which any network, no matter how complicated, can be properly colored with four colors. All the mathematicians who have worked at the problem have come to the conclusion that four colors are probably enough in all cases, so that the latter alternative mentioned above is the more promising.

To see that three colors are not enough, consider the network in Fig. 9:6. Here we have four vertices, and each vertex is joined to each of the other three by an arc. Hence, each vertex must be given a different color, for *each pair* of vertices is joined by an arc. This observation naturally causes us to ask, "Why not draw a network with five vertices, each joined to the other four? Would not such a network require five colors to color it properly?" It certainly would, but this network cannot be drawn *on a plane*. A few trials will convince you that this statement is true. Inasmuch as any network obtained from a map can obviously be drawn on a plane, we must rule out this possibility. (Later on we shall consider networks on surfaces other than planes.)

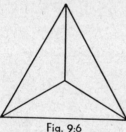

Fig. 9:6

Notice the difference between the network of a map in the above sense and the network consisting of the boundaries of the states of the map. The latter for New England is like that of Fig. 9:7 which is very different from Fig. 9:5. For one thing, Fig. 9:5 has six vertices where Fig. 9:7 has ten. The number of states in the latter is, of course, the number of vertices in the former. In what follows in this

Fig. 9:7

section we consider the map itself and the network consisting of the boundaries in the map.

To indicate how one goes about proving theorems about coloring maps, we shall prove the following theorem:

Any map can be properly colored with six colors.

The proof is quite long, and for convenience we shall break it into several parts.

It will be found useful to consider the entire portion of the plane outside the map as a district of the map. If we can color the map with this added district, we can certainly color it when the district is removed.

A special type of map which we must consider is one in which each vertex is the end-point of exactly three arc-ends. Such a map will be called **regular**. We first prove

1. *It is sufficient to consider only regular maps.* Suppose we have proved (as we shall do later) that every regular map can be colored in six colors. If vertex A is the end-point of only two arcs we can suppress A and unite the two arcs into one, without affecting the coloring of the map. Thus we can consider each vertex to be the end-point of at least three arcs.

We must now show how to eliminate the vertices having more than three arc-ends. Suppose, for example, that there is a vertex with five arc-ends (Fig. 9:8). What we do is construct a new map looking just like the old one except that the part in Fig. 9:8 is replaced by Fig. 9:9. We have then

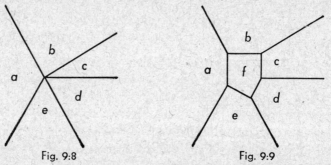

Fig. 9:8   Fig. 9:9

added one new district and replaced the irregular vertex by five regular ones. The same thing can be done for each vertex at which four or more arc-ends come together, and we thus obtain a regular map.

Now we have supposed that this regular map can be colored properly with six colors. In particular, the six districts shown in Fig. 9:9 have colors assigned to them so that no two districts with a common border have the same color.

Now if we assign the same colors to the five regions of Fig. 9:8, this still remains the case, for the abolishment of district f does not introduce any new borders between districts. (Districts like a and c in Fig. 9:8 are not regarded as having a common border.) Since these extra districts can be removed simultaneously over the whole map without materially affecting the remaining districts, we shall have a proper coloring of our original map.

2. *It is sufficient to consider only connected maps.* Suppose we have a map like Fig. 9:10. Here district c is ring-shaped and separates the network into two pieces. However, this is easily avoided as is shown in Fig. 9:11. We have here en-

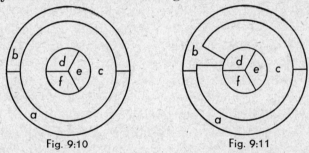

Fig. 9:10   Fig. 9:11

larged district b at the expense of c, thereby connecting the two parts of the network. Obviously Fig. 9:10 can be colored with six colors if Fig. 9:11 can, and as this process works in any case of this sort, we may suppose that our network is connected.

The next thing we prove is a theorem about networks which apparently has nothing to do with maps. However, the connection will be evident later.

3. *If a connected network on a plane has V vertices and E arcs, and if it separates the plane into F districts, then $F + V - E = 2$.* This is called **Euler's theorem.**

Any connected network can be built up, or drawn, an arc at a time, each new arc being attached at one or both ends to the part already drawn. Consider the effect on the expression $F + V - E$ of adding such an arc to the network. If the arc is attached at only one end, we do not change the number of districts, but we increase the number of arcs by one, and also the number of vertices by one, for we get a new vertex at the free end of the arc. That is, F is unchanged and V and E are each increased by one, so that $F + V - E$ remains the same. If we put a vertex on an arc, we increase V by 1 and E by 1 and do not change $F + V - E$. If both ends of the arc are attached at vertices to the part of the network already drawn, we do not change V, but E is increased by one and so is F, for the addition of such an arc evidently separates one of the previous districts into two. Hence, in this case also, $F + V - E$ is not changed. Now when we start drawing the network we have a single arc, with two vertices, and one district, so that $F + V - E = 1 + 2 - 1 = 2$. Since the value of $F + V - E$ is not changed by adding arcs, it is still equal to 2 for the complete network.

One figure which at first glance appears to violate this result is the circumference of a circle which has no vertex, one arc, and divides the plane into two regions, but such a figure does not satisfy our definition of a network since it has less than two vertices. If we put two vertices on the circumference, Euler's theorem holds. (As it happens, the theorem still holds if just one vertex is put on the circumference, though in this case we have no network.)

This property of networks on a plane is very useful in many connections, and we shall have occasion to refer to it later.

4. *A regular connected map has at least one district which touches five or fewer others.*

Consider the network whose vertices are the points where three of the districts meet and whose arcs are the boundaries between the districts. Using V, E, F as in 3, we have

(A) $$F + V - E = 2,$$

provided this network is connected.

Let n_2, n_3, $\cdots$ denote the number of districts of the map which have 2, 3, $\cdots$ sides, that is, which touch 2, 3, $\cdots$ other districts. A district with k sides has k arcs on its boundary, hence the n_k such districts have kn_k arcs on their boundaries, so the districts have in all $2\,n_2 + 3\,n_3 + 4\,n_4 + \cdots$ arcs on their boundaries. But each arc is on the boundary of two districts, and so

(B) $\qquad\qquad 2\,n_2 + 3\,n_3 + 4\,n_4 + \cdots = 2\,E.$

Now consider the relationship between arcs and vertices. Each of the E arcs has two ends and there are three of these $2\,E$ ends at each of the V vertices since the map is regular. Hence

(C) $\qquad\qquad\qquad 2\,E = 3\,V.$

Finally

(D) $\qquad\qquad\qquad n_2 + n_3 + n_4 + \cdots = F.$

Multiply equation (A) by 6 to get $6\,F + 6\,V - 6\,E = 12$. Since equation (C) implies $4\,E = 6\,V$, we have

$$6\,F + 6\,V - 6\,E = 6\,F + 4\,E - 6\,E = 6\,F - 2\,E = 12,$$

and, substituting in this the values of F and $2\,E$ from equations (D) and (B), we obtain

(E) $\quad (6-2)n_2 + (6-3)n_3 + (6-4)n_4 + (6-5)n_5$
$\qquad\qquad + (6-6)n_6 + (6-7)n_7 + \cdots = 12.$

Now we are trying to prove that there is at least one district with five or fewer sides. If this were false, we should have $n_2 = n_3 = n_4 = n_5 = 0$ and equation (E) would say that

$$(6-6)n_6 + (6-7)n_7 + \cdots = 12.$$

But this is impossible, for the first term is zero and all the rest are either zero or negative, and so their sum cannot be positive. Hence n_2, n_3, n_4, n_5 cannot all be zero; this proves statement 4.

Note that up to this point we have made no use of our six colors. Everything we have said so far will apply to the problem of four colors, or five colors.

5. *Every regular map can be properly colored with six colors.*

We have just proved that a regular map has a district with five or fewer sides. Suppose there is a five-sided district (Fig. 9:12). Consider the map obtained by removing the boundary between this district and one of the adjacent ones (Fig. 9:13). If this new map can be colored with six colors, so can the old one, for since the district a touches only five others, there is surely a color available for it when it is put back. The same argument will obviously work for a district with less than five sides.

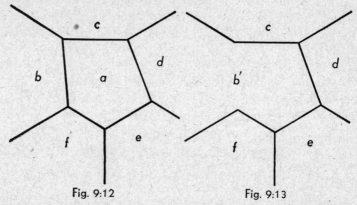

Fig. 9:12   Fig. 9:13

So the question now is, "Can the new map, obtained by coalescing districts a and b, be colored?" Well, this new map is still regular (after we eliminate the two vertices common to just two arcs), and so the same argument can be applied to it. Continuing in this way we can reduce the number of districts until we have only six left. We color these all differently, and then work backward, reintroducing the suppressed districts one at a time, until we have colored the original map.

There is one possible complication in this process. If a district is coalesced with one which touches it in two places, a ring-shaped district is created. Before proceeding any farther we must first eliminate this ring by the process indicated in the proof of statement 2.

Combining statements 5 and 1 we have a proof of the six-color theorem.

The six colors occur only in proving statement 5, and it is easy to see why we may need six. As a matter of fact, we shall, in the next section, by a somewhat longer argument, prove that even with only five colors the coalescing process can still be carried out, and thus prove the five-color theorem. But nobody has yet been able to push the process one step farther. The best result that has been obtained so far is that any map with less than thirty-six districts can be properly colored with four colors.

For certain special kinds of maps, less than four colors are needed. A checkerboard is an example of a map requiring only two colors. It can be shown that two colors are sufficient if the network formed by the boundaries of the districts has no odd vertices. A little experimenting with such maps will enable you to see why this is so.

4. The five-color theorem.

For those interested we now show the modification of the above argument necessary to prove that every map may be colored with five colors. Parts 1 to 4 of the previous argument hold as before and we have given a map containing a five-sided district as in Fig. 9:11 and wish to show: if a map obtained by eliminating one or more boundaries of the given map can be colored with five colors, then so can the given map.

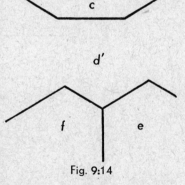

Fig. 9:14

In this case we erase the boundary common to districts a and d as well as that common to districts b and a and have Fig. 9:14. Then regions a, b, and d become one district d' of the new map. Suppose this map may be colored with five colors. Then the color of d' may be allotted to b and d of the given map, leaving a fifth color for district a, *unless*, in the given map, b and d had a boundary in common. If, however, b and d have a common boundary, the three districts a, b, and d completely surround either district c or districts f and e. In both cases it will be impossible for c to have a boundary in common with either f or e. Then we may go back to Fig. 9:11 and erase the boundaries between a and c as well as a and f and carry through our argument as above.

EXERCISES

1. Draw the network which represents the states of the United States in the same way that Fig. 9:5 does the New England states. Use the network to answer the questions.

    a. What is the smallest number of states that must be passed through traveling from Portland, Maine, to Los Angeles, never going outside the country?

    b. How does this compare with the least number of states traversed in going from Miami to Seattle?

    c. Can the map of this country be colored with three colors? If not, where are the exceptional portions?

    d. In how many ways can one remove one state from the country to leave the remaining network disconnected?

    e. What states or state border on the greatest number of other states where "border on" means have a common boundary which is more than a point?

2. Color with four colors the map in Fig. 9:15.

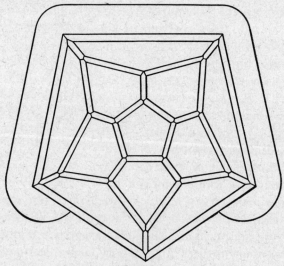

Fig. 9:15

3. Draw a regular map with ten districts that can be colored with three colors, no two adjacent districts having the same color.

4. Find all the relationships you can between the F, V, and E for Fig. 9:5 and the F, V, and E for Fig. 9:7.

5. If all places outside New England are considered part of one vast region (inhabited only by savages), by what should Fig. 9:5 be replaced? Then discuss the questions raised in Exercise 4.

6. Given any connected network A and a network B obtained from A by erasing one of the arcs of A. (It is understood that if the erasure leaves an isolated vertex, the vertex also is erased.) Show that if F, V, and E are the numbers of faces, vertices, and arcs, respectively, of network B, then $F + V - E = 3$ if B is not connected.

* 7. What are the possible values of $F + V - E$ for a network obtained by erasing two of the arcs of a connected network?

5. Topology.

Let us stop at this point to consider the kind of problems we have been discussing in connection with networks. Networks are geometrical objects; that is, they are composed of points and lines, and yet we have used none of the usual geometrical methods in dealing with them. We have not been concerned, for instance, with the distances between the vertices, or the angles between the arcs, or the areas of the regions in the networks. Our only concern was the way in which the various arcs were hooked together at the vertices. The network arising from the New England states could have been drawn in the form of Fig. 9:16 just as well as in the form of Fig. 9:5.

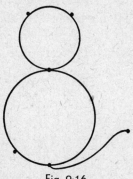

Fig. 9:16

In our plane geometry in high school we dealt with properties of figures which remained unchanged if we moved them from place to place without changing their shape. A triangle formed of three sticks has the same area whether it is in New York or Timbuktu. The lengths of its altitudes and its medians are left unaltered by such a journey if the sticks remain the same size. Two figures were called congruent if by **rigid motion,** that is, moving without distortion, we could make one coincide with the other. As far as our plane geometry was concerned, two congruent figures were identical: they had equal angles, equal sides — everything that we investigated in plane geometry was the same for one as for the other.

In **topology** we allow ourselves more freedom of action. In addition to rigid motions we allow twisting, stretching, and bending, but not cutting or welding. We allow ourselves to bend crowbars into U's but not to weld them into hoops. We can make circular hoops into oval hoops or even into the outlines of a square, but not into crowbars. Such a transformation we call a **topologic transformation,** as opposed to the transformation of rigid motion which was fundamental in plane geometry. Just as in plane geometry we called two figures **congruent** if we could make them coincide by rigid motion, so in topology we call figures **equivalent** if we can make them coincide by a topologic transformation. Just as in plane geometry we investigated the properties left unchanged by rigid motions, so here in topology we are interested in properties left unchanged by topologic transformations. The traversing of a path and the coloring of a map are topological problems because the shape of the path or the shape of the boundaries of a map make no difference to him who wishes a system for traversing the path or coloring the map. On the other hand we have now no concern with angles, areas, and such things which can easily be changed by a little twisting.

It must be emphasized that a topologic transformation never brings together two (or more) points that were originally separate, nor does it tear a figure apart so that one point becomes two. Thus Figures 9:17 are not obtainable from Fig. 9:16 by a topologic transformation. It is evident that these figures do not represent the New England states.

While our definition of a topological transformation allows us to round off sharp corners, from some points of view it would not allow us to shake off vertices. However, it is con-

venient to agree that, before considering equivalence of two
figures, we shall *eliminate all vertices which are the ends of
exactly two arcs.* For instance, we should say that a line
segment with ten points marked off on it would be equivalent to a line segment with no points marked on it; the
perimeter of a square with vertices at its corners and the

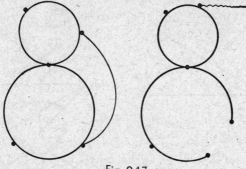

Fig. 9:17

midpoints of its sides would be equivalent to such a perimeter after the vertices are eliminated, that is, would be
equivalent to the circumference of a circle. That a circle
without vertices is not a network is here immaterial since
it is the *figure* with which we are concerned. In fact, we
may apply this definition of topological equivalence to three-
dimensional figures and say, for instance, that the surface
of a sphere is equivalent to that of a cube, a pyramid, or a
prism. On the other hand, a circle (that is, the area bounded
by the circumference) is not equivalent to the ring-shaped
area between two concentric circles nor is a sphere equivalent to an inner tube.

EXERCISES

1. Pair off the following figures so that the two figures in each pair
are equivalent.
    a. The circumference of a circle.
    b. A line segment.
    c. The surface of a sphere.
    d. A hollow sphere.
    e. A solid sphere.

  f. The surface of a cube.

  g. A solid cube.

  h. A cube with a hole bored through it.

  i. The surface of *h.*

  j. The area between two concentric circles.

  k. The surface of a doughnut.

  l. The edges of a tetrahedron.

  m. A piece of gas pipe.

  n. A piece of gas pipe with the ends plugged.

  o. A swastika.

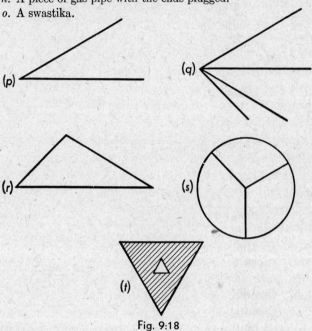

Fig. 9:18

2. The following newspaper comment appeared after a popular lecture on topology: "Topology is the branch of mathematics which tells us how to make inner tubes out of billiard balls." What are your comments on the comment?

6. Planar networks.

To illustrate some applications of these ideas let us consider in detail a point which arose in connection with the four-color problem. It was stated that the networks arising from maps were such that they could be drawn in a plane.

However, the conclusions which one reaches by considering the topological properties of such networks apply to any network which is merely *equivalent* to one in a plane. A network equivalent to a network in a plane, we call a **planar network.** Thus the twelve edges and eight vertices of a cube form a network which does not lie in a plane, but which is equivalent to the planar network of Fig. 9:19, where one face of the cube corresponds to the portion of the plane outside the network. Hence the edges and vertices of a cube form a planar network and the conclusions reached in discussing the four-color problem apply to the network on the cube.

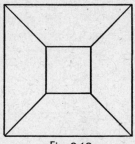

Fig. 9:19

The most important of these conclusions is that given in statement 3, namely, that $F + V - E = 2$. For the cube this gives us $6 + 8 - 12 = 2$. This relation between Faces, Vertices, and Edges, therefore, applies to any polyhedron (solid bounded by planes) which is equivalent to a sphere.

An important problem in topology is the characterization of planar networks. It was stated that the network with five vertices, each joined to the other four, is not planar. Another network which is not planar arises from the old problem: There are three houses, A, B, C, and three wells, X, Y, Z. The owners of the houses each wish to construct a path to each of the three wells, but being very suspicious of each other they will not permit two paths to cross. How can the paths be made? Unless you allow one of the paths to go directly over a well, or through one of the houses, the problem has no solution. This is because the network obtained by joining each of three points, A, B, C, to each of three others, X, Y, Z, is not planar.

To determine whether or not a complicated network is planar is not an easy matter. Consider, for example, the network consisting of the edges of a cube plus the lines joining the center to each of the vertices. (Figure 9:20

represents an equivalent network if we assume that the arcs AI, BI, $\cdots$ hop over EF, FG, $\cdots$ without intersecting them.) To see that this graph is not planar consider the portion of it drawn in Fig. 9:21. Here we have five vertices joined in all possible ways by arcs, and we have said that

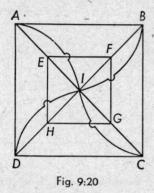

Fig. 9:20

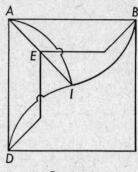

Fig. 9:21

such a network is not planar. Hence the complete network cannot be planar either.

This method can be applied to any network. If we can find a portion of the network which is equivalent either to five points joined in all possible ways, or to three points each joined to each of three others, then our network is not planar. It is interesting that the test also works in the other direction. If no such portion of the network can be found, then the network is planar. This is very difficult to prove.

EXERCISES

1. Draw networks in a plane equivalent to:
    a. The edges of a square pyramid,
    b. The edges of a regular octahedron.
2. Show on the basis of statements made above that the network consisting of the edges of a cube and one diagonal of the cube is not planar.
3. Which of the following networks are planar? Show that all non-planar ones check with the statements of the paragraph preceding these exercises.
    a. The edges of a tetrahedron and one of its altitudes.
    b. The edges of a box and the diagonals of its faces.

c. The network whose vertices correspond to the faces of a cube and where any edge common to two faces corresponds to a line connecting the vertices of the network.

d. The edges of a tetrahedron, a point within it, and the lines connecting the point with the vertices of the tetrahedron.

e. The network of Fig. 9:6 with the addition of one vertex, *E*, outside the triangle, and an arc connecting *E* with the vertex inside the triangle.

7. Surfaces.

Probably the simplest surface which is encountered in elementary geometry is the surface of a sphere. This is defined as the locus of points at a given distance from a fixed point, called the **center.** In solid geometry many things are proved about such surfaces, including properties of tangent planes, or great and small circles on the surface, and various others. None of these properties, however, is of any interest from the point of view of topology, for if the surface is stretched or deformed in some irregular manner the very notion of tangent plane or great circle may become meaningless. If we think of the spherical surface as a rubber balloon, the topological properties of the surface are the ones which are not altered when the balloon is stretched, pinched, or deformed in any manner *without tearing.* Since by such a process we can stretch the balloon into the shape of the surface of a cube, or of any polyhedron, it is difficult at first to conceive of any properties of the surface which would not be changed by such stretching. However, there are such properties. For instance, the balloon has an inside and an outside, and no amount of stretching or twisting will alter this property. Hence, this is a topological property of a spherical surface. This property appears so obvious that in elementary solid geometry we just take it for granted without even commenting on it. This is true of a great many topological properties, and it is probably the main reason why topology is such a recent development in mathematics.

Another topological property of a spherical surface is the fact that any closed curve drawn on the surface divides it into two parts. By a closed curve we mean any line, curved or partly straight, whose ends are joined but which does not cross itself. Thus Figs. 9:22a and 9:22b are closed curves,

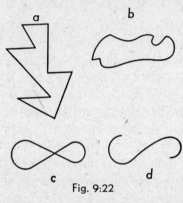

but Figs. 9:22c and 9:22d are not. Or, in other words, a closed curve is any figure topologically equivalent to the circumference of a circle. This property of a spherical surface is evidently a topological one, and, like the other, seems to be obviously true. However, there are surfaces which do not have this property. The simplest of these is the *torus*,

Fig. 9:22

which is the surface of a doughnut, or an inner tube. In Fig. 9:23 there have been drawn on such a surface two closed curves, neither of which divides the torus into two parts. As a matter of fact both curves together do not separate the surface, for it is easily seen that we can pass

from any point of the surface to any other without having to cross either curve. However, it is fairly evident that no more closed curves can be drawn without spoiling this property.

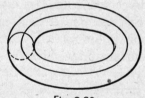

These observations concerning closed curves lead us to the conclu-

Fig. 9:23

sion that a torus is not equivalent to a spherical surface. In more familiar terms, a spherical balloon cannot be deformed into the shape of an inner tube. It is not difficult to see how we can get surfaces which are not equivalent to either the spherical surface or the torus. Two such surfaces are buttons with two and three holes respectively. Obviously we can construct such examples with as many holes as we wish. It can be proved that on the surface of a button with n holes there can be drawn precisely $2n$ closed curves without separating the surface into two or more parts. Hence two buttons with a different number of holes are not topologically equivalent, whereas two with the same number of holes are equivalent. This gives us a complete topological classification of buttons. Of course, the classification extends not only to objects in the shape of a button, but to any object whatever. Thus the surface of one of the chairs in your classroom is equivalent to the surface of a button with holes. (How many?)

So far we have used the word "surface" to refer to the complete surface of some solid object. Let us now use this word for such things as a rectangle, or a circle, or the lateral area of a cylinder or a cone. That is, we consider surfaces with *edges*. In contrast to this usage of the word, we shall refer to the kind of surfaces we considered previously as

Fig. 9:24

closed surfaces. Surfaces with edges may also be analyzed by considering the curves which can be drawn on them. For example, a rectangle is divided into two parts by any curve joining two points of its edge, whereas on a cylindrical surface one such curve may be drawn without separating the surface. Hence a rectangle and a cylindrical surface are not equivalent.

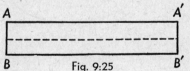

Fig. 9:25

A cylindrical surface may be obtained by gluing together the ends of a rectangular strip, so that A coincides with A' and B with B' in Fig. 9:25. An interesting surface, known as a **Möbius strip,** is obtained by gluing the ends so that A coincides with B' and B with A' (Fig. 9:26). This surface has some surprising properties. Like the cylinder we can cut it on a line joining two points of the edge without separating the surface, and yet the resulting edge consists of a single closed curve. Unlike the cylinder, we can cut the surface along a closed curve (the dotted line in the figure) and still have a single piece left. But perhaps the most remarkable property is the following. Imagine an ant crawling along the surface but never crossing the edge. If she starts at point C and follows the dotted line, she will eventually return to point C, *but on the other side of the surface.*

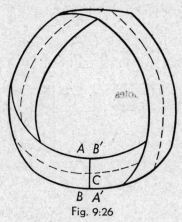

In other words, this surface has the property that you can get from any point of it to any other without crossing over an edge. This is certainly not the case with any closed surface, nor with the other open surfaces we have considered. Surfaces behaving like the Möbius strip in this respect are said to be "one-sided."

A B'

B A'

Fig. 9:26

Returning to Fig. 9:25, let us again glue A to A' and B to B', but giving the line $A'B'$ a twist through 360° before gluing. We then get a strip which, although twisted, has most of the properties of the untwisted cylindrical surface, at least as far as the behavior of curves on it is concerned. If we cut this strip along the center line, we obtain two pieces as we should also for the cylinder, but unlike the case of the cylinder the two pieces are linked together so that it is impossible to separate them. Other interesting surfaces can be made by giving the strip several twists before gluing the ends together.

For brevity we will use the term "n-strip" to refer to the surface obtained by twisting the strip in Fig. 9:25 through $n \cdot 180°$ before gluing the ends together. Thus a cylinder is a 0-strip, and a Möbius strip is a 1-strip.

EXERCISES

1. Show that an n-strip is one-sided if n is odd and two-sided if n is even.

2. Show that if an n-strip is cut down the middle (that is, as in Fig. 9:26) we obtain one piece if n is odd and two pieces if n is even.

3. If a 1-strip is cut down the middle, we obtain a single piece which is an n-strip. Why is this so, what is the value of n, and is the n-strip one- or two-sided? If this is cut again down the middle, what happens?

4. If a 2-strip is cut down the middle, each of the two pieces is a two-sided n-strip. Why is this so, and what is the value of n? If each of these strips is cut again down the middle, what happens?

5. A torus can be made from the strip in Fig. 9:25 by gluing edge AA' to edge BB' to make a hollow cylinder and then gluing the ends together to form the torus. Show the two curves on the torus of Fig. 9:23 as they would appear on the strip of Fig. 9:25. (Physically this would be hard to manage with paper but not with rubber.)

6. Show how to draw on a button with two holes, four lines which do not separate the surface.

8. Maps on surfaces.

In discussing the four-color problem we considered maps drawn on a plane. It is evident that the same problem can be investigated for maps on a sphere, a torus, or any other surface. As might be expected, maps on different kinds of surfaces require a different number of colors, and the more complicated the surface the more colors are required.

It is not hard to see that a map on a sphere will require just as many colors as a map on a plane. For suppose we have a map on a sphere. Thinking of the spherical surface as a rubber balloon, cut a small hole in one of the districts. By pulling out the circumference of this hole, we can evidently stretch the balloon until it becomes flat. We then have a map on a plane which is equivalent to the original one of the sphere, and hence the same number of colors will be required for each of them.

For the next simplest closed surface, the torus, the situation is quite different. Figure 9:27 is an example of a map consisting of seven districts, each one touching all the remaining six. Such a map will obviously require seven colors. Now the interesting thing about the torus is that we can prove that seven colors will suffice in all cases. The proof is very much like the one outlined for six colors in the case of the plane. By the same methods that we used for maps on a plane we can show that any map on a torus must have at least one district which touches six or fewer others. Since we have seven colors at our disposal, this district can be coalesced with one of the adjacent ones, and the process

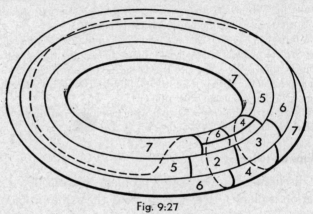

Fig. 9:27

continued until there are only seven districts left. Hence the map-coloring problem for the torus is completely solved: seven colors is the minimum number.

In the same way it has been proved that for the surface of a button with 2, 3, or 4 holes the number of colors needed is 8, 9, and 10 respectively. It is rather a strange state of affairs that these more complicated surfaces are so easy to deal with, while the much simpler plane and sphere cause so many difficulties.

The relationship between maps and networks can, of course, be extended to maps on the torus or any other surface. We saw that on a plane — and hence also on a sphere — we could not have a network joining five points in all possible pairs. On a torus, however, we can join as many as seven points in all possible ways. This possibility is assured by the existence of the map in Fig. 9:27. The fact that no map on a torus requires more than seven colors tells us that a network joining eight points in all possible ways cannot be drawn on a torus. Similar statements can be made for the other types of closed surfaces.

A convenient way of dealing with the drawing of a map on the torus is by means of a figure like Fig. 9:28. If we

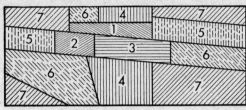

Fig. 9:28

cut a sheet of rubber in the form of the figure and divide it into districts as indicated, then we may (1) roll it into a tube by making the top and bottom edges coincide, and (2) stretch and twist the tube until its ends coincide as indicated in Exercise 5 of the last section. As a result we will have essentially the torus of Fig. 9:27.

EXERCISES

1. Draw a picture of a network on a torus which consists of 7 vertices and all the arcs joining them in pairs.

2. Show that the problem of coloring a map on a cylinder is the same as for a sphere or a plane.

3. Construct a map on a Möbius strip which consists of 6 districts each touching the other 5. (*Hint:* Draw the map on the rectangle of Fig. 9:25, remembering how the ends are to be glued together.)

4. Find $F + V - E$ for the network on the surface of the torus in Fig. 9:23; for the curves drawn on the button with 2 holes in Exercise 6 of the previous section.

9. Topics for further study.

Most of the topics dealt with in this chapter are touched upon in Chap. 8 of reference **25**. For a more complete account of the four-color problem and an indication of the proof for five colors see reference **16**. See reference **5**, Chaps. 8, 9.

Bibliography

1. Abbott, E. A., *Flatland, a Romance of Many Dimensions*, Little, Brown & Company, Boston, 1929. This is a popularly written account of life in a world of other than three dimensions.
2. Andrews, F. E., "An Excursion in Numbers," *The Atlantic Monthly*, Vol. 154, 1934, pp. 459–466. This and the following book describe to non-mathematicians the wonders of a number system to the base 12.
3. ——, *New Numbers*, 2d ed., Essential Books, New York, 1944.
4. ——, "Revolving Numbers," *The Atlantic Monthly*, Vol. 155, 1935, pp. 208–211. This is a very interesting account for non-mathematicians of cyclic or "revolving" numbers. The author's results are stated by means of examples with no attempt at proof.
5. Ball, W. W. R., *Mathematical Recreations and Essays*, 11th ed., The Macmillan Company, New York, 1939. This is the most widely consulted book on mathematical puzzles. In many cases rather complete theories are developed.
6. Bell, E. T., *Men of Mathematics*, Simon and Schuster, Inc., New York, 1937. This book contains a set of interestingly written short biographies of famous mathematicians and describes the ideas they originated.
7. ——, *Numerology*, The Williams & Wilkins Company, Baltimore, 1933. This details a mathematician's opinion of one of the occult pseudo sciences.
8. Cooley, Gans, Kline, and Wahlert, *Introduction to Mathematics*, Houghton Mifflin Company, Boston, 1937. This is a text in Freshman mathematics.
9. Dantzig, Tobias, *Number, the Language of Science*, The Macmillan Company, New York, 1930. This is an interesting, popularly written account of the history of the number con-

cept and number systems, number lore, and the theory of numbers.

10. Davis, H. T., *College Algebra*, Prentice-Hall, Inc., New York, 1940. This text contains some historical material and other topics of interest not usually included in a book on this subject.

11. Dresden, Arnold, *An Invitation to Mathematics*, Henry Holt and Company, Inc., New York, 1936. The capable student can find in this book a more complete treatment than we can give here of many of the topics discussed.

12. Dudeney, H. E., *Amusements in Mathematics*, Thomas Nelson & Sons, London, 1917. This book and the three following are excellent collections of puzzles. Quite often the author raises a question which he does not answer and which can serve as a starting point for an investigation by an interested student.

13. ——, *Canterbury Puzzles*, E. P. Dutton & Company, Inc., New York, 1908.

14. ——, *Modern Puzzles*, C. Arthur Pearson, London, 1926.

15. ——, *Puzzles and Curious Problems*, revision by James Travers, Thomas Nelson & Sons, London, 1931.

16. Franklin, Philip, "The Four Color Problem," *Scripta Mathematica*, Vol. 6, 1939, pp. 149–156, 197–210.

17. Gale and Watkeys, *Elementary Functions*, Henry Holt and Company, Inc., New York, 1920. An elementary text containing an especially good introduction to statistical graphs.

18. Ginsburg, Jekuthial, "Gauss's Arithmetization of the 8-Queens Problem," *Scripta Mathematica*, Vol. 5, 1938, pp. 63–66.

19. Griffin, F. L., *Introduction to Mathematical Analysis*, Houghton Mifflin Company, Boston, 1921. An elementary text.

20. Guttman, Solomon, "Cyclic Numbers," *The American Mathematical Monthly*, Vol. 41, 1934, pp. 159–166. The first part of this article should be understood by one who knows only a little algebra.

21. Hardy, G. H., *A Mathematician's Apology*, Cambridge University Press, Cambridge, 1940. A fascinating account of a mathematician's outlook on life, written by one of the finest mathematicians of the present day.

22. Hilbert, David, *The Foundations of Geometry*, trans. by E. J. Townsend, The Open Court Publishing Company, La Salle, Ill., 1902.

23. Hogben, L. T., *Mathematics for the Million*, George Allen & Unwin, London, 1936. This was written by a British economist during a long convalescence — written with the object of informing his nonmathematician friends about mathematics. It has been a very widely sold and read book.

24. Johnson, W. W., "Octonary Numeration," *The New York Mathematical Society Bulletin*, Vol. 1, pp. 1–6.

25. Kasner and Newman, *Mathematics and the Imagination*, Simon and Schuster, Inc., New York, 1940. This book explores in a fascinating way the beginnings of many of the important mathematical fields of today.

26. Kraitchik, Maurice, *Mathematical Recreations*, W. W. Norton & Company, Inc., New York, 1942. This book contains material similar to that in Ball's book referred to above.

27. Lieber, L. R. and H. G., *Non-Euclidean Geometry*, The Science Press Printing Company, Lancaster, Pa., 1931. An amusingly written and illustrated account of a geometry different from that in high-school textbooks.

28. Merrill, H. A., *Mathematical Excursions*, Bruce Humphries, Inc., Boston, 1934. This book explores some of the bypaths which the author and her students have found interesting. Most high-school Seniors could understand it.

29. Merriman, G. M., *To Discover Mathematics*, John Wiley and Sons, Inc., New York, 1942. An unusual elementary text written in essay style.

30. Quine, W. V., *Elementary Logic*, Ginn and Company, Boston, 1941. An elementary account of the basis of logic from the modern point of view.

31. Richardson, Moses, *Fundamentals of Mathematics*, The Macmillan Company, New York, 1941. This is a college Freshman text which is worth exploring.

32. Smith, D. E., *The History of Mathematics*, Vol. 2, Ginn and Company, Boston, 1925. This book gives a rather complete and authoritative history of mathematics; here references may be found to still more complete works.

33. Terry, G. S., *Duodecimal Arithmetic*, Longmans Green and Company, New York, 1938. An ardent disciple of F. E. Andrews (see references above) here publishes various tables for the number system to the base twelve.

34. Tingley, E. M., "Base Eight Arithmetic and Money," *School Science and Mathematics*, Vol. 40, 1940, pp. 503–508.

35. Uspensky and Heaslet, *Elementary Number Theory*, McGraw-Hill Book Company, Inc., New York, 1939. In many places this book will be too difficult for the college Freshman, but there are parts of it which can be understood without much mathematical experience and background. It is a very original and thought-provoking book.

36. Veblen, Oswald, "The Foundations of Geometry," in *Monographs on Topics of Modern Mathematics* by J. W. A. Young, Longmans Green and Company, New York, 1911. Some of the axioms neglected by Euclid are described here.

37. Weyl, Hermann, "Emmy Noether," *Scripta Mathematica*, Vol. 3, 1935, pp. 201–220. This is a beautifully written story of the life and works of the most famous woman mathematician of modern times. Anyone who reads it should glean some feeling for what it is that goes to make a great mathematician.

38. Witherspoon, J. T., "A Numerical Adventure," *Esquire*, December, 1935, pp. 83ff. An account of casting out the nines and related phenomena.

Index